The Quantum Frontier

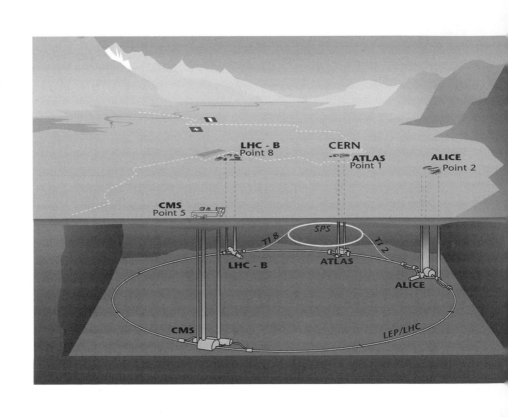

The Quantum Frontier

The Large Hadron Collider

Don Lincoln

Foreword by Leon Lederman

The Johns Hopkins University Press
Baltimore

© 2009 The Johns Hopkins University Press
All rights reserved. Published 2009
Printed in the United States of America on acid-free paper
9 8 7 6 5 4 3 2 1

The Johns Hopkins University Press
2715 North Charles Street
Baltimore, Maryland 21218-4363
www.press.jhu.edu

Library of Congress Cataloging-in-Publication Data

Lincoln, Don.
    The quantum frontier: the large hadron collider / Don Lincoln; foreword by
    Leon Lederman.
        p. cm.
    Includes bibliographical references and index.
    ISBN-13: 978-0-8018-9144-1 (hardcover: alk. paper)
    ISBN-10: 0-8018-9144-2 (hardcover: alk. paper)
    1. Higgs bosons. 2. Large Hadron Collider (France and Switzerland).
3. Particles (Nuclear physics). I. Title.
QC793.5.B62L56 2009
539.7'376—dc22   2008022647

A catalog record for this book is available from the British Library.

*Special discounts are available for bulk purchases of this book. For more information,
please contact Special Sales at 410-516-6936 or specialsales@press.jhu.edu.*

The Johns Hopkins University Press uses environmentally friendly book
materials, including recycled text paper that is composed of at least 30 percent
post-consumer waste, whenever possible. All of our book papers are acid-free,
and our jackets and covers are printed on paper with recycled content.

To those giants on whose shoulders I have stood

Contents

Foreword

The Large Hadron Collider, or LHC, is a new scientific tool. The invention of tools, instruments to aid in observation and measurement, has been crucial to the advancement of science. Even though there is a robust debate as to the relative virtues of pure versus applied research, instruments are vital to both branches and serve as a harmonious bridge. In the late nineteenth and early twentieth centuries, progress in both basic research and applied research has been utilized to create ever more powerful tools. Many of these were designed for comfort and entertainment but their use to advance the understanding of nature led the way. It's really cozy: research creates new knowledge, which enables the creation of new instruments, which make possible the discovery of new knowledge.

An example: Galileo constructed many telescopes after hearing about their invention in Holland. In one stunning weekend, he turned a telescope to the sky and discovered four of the moons of Jupiter! This convinced him that indeed the Earth was in motion as surmised by Copernicus. The evolution of telescopes ultimately gave humans a measure of the vastness of our universe with its billions of galaxies, each hosting billions of suns. And in the more sophisticated science, more powerful telescopes were developed.

A further example relevant to our book about the LHC: the structure and properties of electrons are about as basic as one can get in the grand quest for understanding how the world works. But many of these properties make electrons a powerful component in countless instruments. Electrons make x-rays for medical use and for determining the structure of biological molecules. Electron beams make oscilloscopes, televisions, and hundreds of devices found in laboratories, hospitals, and the home.

An impressive technology enabled the control of energetic electron beams in particle accelerators. These were invented in the 1930s and provided precise data on the size, shape, and structure of atoms. To probe the nucleus of atoms, higher energies were required, and the acceleration of protons was added to the toolkit of physicists.

An approximate timetable of progress in accelerators may be useful and is shown below. Note that eV equals one electron volt, so keV is 10^3 electron volts, MeV is 10^6 electron volts, GeV is 10^9 electron volts, and TeV is 10^{12} electron volts. You can see in the table that the higher the energy of the accelerated particle, the smaller the distance probed. However, to probe the very small, the accelerators also grew in size, complexity, and cost. Accelerators are then in essence powerful microscopes, taking over when light is no longer sufficient.

Date	Energy	Distance Probed
1930	~100 keV	10^{-11} meters
1950	~100 MeV	10^{-14} meters
1970	100 GeV	10^{-17} meters
1990	1 TeV	10^{-18} meters
2010	10 TeV	1^{-19} meters
2020	?	?

Over the past 80 years, hundreds of accelerators have been constructed worldwide, predominantly to address the unknowns in the field of particle physics. Other applications of accelerators are these: in medical treatment, as powerful x-ray sources, in industry, and in oil explorations. The complexity and cost of the newer machines have forced large international collaborations. For the first time, construction costs of an accelerator, the LHC, will be shared by Europe, Russia, Japan, China, and the United States.

There is a matching set of requirements for the construction of the detectors (see chapter 4) that must observe the new domain exposed by the accelerators—essentially supermicroscopes. Here, intimate collaborations of over a thousand scientists and students are involved. The official language of these collaborations is, of necessity, "broken English."

It should be noted that, though high energy physics came out of a marriage of nuclear and cosmic ray physics in the late 1940s, we now recognize a new merger of high energy particle physics, which is accelerator based, with astrophysics, which is telescope based. The long-recognized connections of the inner space of particles with the outer space of the cosmos has been reinforced by baffling data on gravitation (dark matter and dark energy) and the continuing mystery of particle symmetry-breaking. However, the "inner space-outer space" connection teaches us that the newly born universe consisted of the elementary particles out of which the stars, galaxies, planets, and people eventually emerged.

So, in the first decade of the twenty-first century, the venerable Tevatron accelerator at Fermilab, born in the scientific dreams of 1985, is operating at full capacity in the hopes of adding to its distinguished list of discoveries before the advent of its CERN (in Geneva, Switzerland—the lab we love to hate) successor, the LHC, scheduled to begin operations in 2008.

At the entrance to the accelerator, the atmosphere is heavy with the promise of discovery. The list of burning open questions today is longer and more profound than that with which we struggled in 1985 (see chapter 5 for a few of today's questions).

Our list of questions will not all be solved by the LHC, and new ones will surely be added. For now, a new generation of accelerators grows in the minds and in the R & D of a new generation of accelerator physicists and their students.

This is a glorious time for them.

But in the meantime, this book by Don Lincoln tells of the excitement experienced by physicists as the LHC commences operations and lets the reader appreciate why the LHC is of such great interest to all physicists. We live in very interesting times.

Leon Lederman

A few quotes as salsa for the repast that awaits you in the journey ahead with Don Lincoln.

One of man's enduring hopes has been to find a few simple general laws that would explain why nature, with all its seeming complexity and variety, is the way it is.

We will still need the LHC to pin down the details of the symmetry-breaking mechanism that gives mass to elementary particles.

Steve Weinberg, Nobel laureate, Physics 1979

The supreme test of the physicist is to arrive at those universal elementary laws from which the cosmos can be built up by pure deduction.

Albert Einstein

When Anton von Leeuwenhoek first saw his "animacules" in a drop of pond water in the seventeenth century, he was in fact extending the ability of humans to see the world in modes not accessible to eyes alone.

The number of dimensions is the number of quantities you need to know to completely pin down a point in space.

Supersymmetry is an extension of known particle physics concepts and has a good chance of being tested in forthcoming experiments. String theory is different.

Lisa Randall, professor of physics, Harvard University

The expanding cloud of billions of galaxies that we call the Big Bang may be just a fragment of a much larger universe in which Big Bangs go all the time, each with different values for the fundamental constants.

Andrei Linde, professor of physics, Stanford University

Every day in a handful of particle accelerators throughout the world, scientists accelerate protons or electrons to tremendous energies and collide them. In these collisions it is possible to create, for a brief instant, the conditions that have not existed in the universe for fourteen billion years.

Edward "Rocky" Kolb, professor of astrophysics,
University of Chicago

The scientist does not study nature because it is useful to do so. He studies it because he takes pleasure in it and he takes pleasure in it because it is beautiful. If nature were not beautiful, it would not be worth knowing and life would not be worth living.

It is because simplicity and vastness are both beautiful that we seek simple facts and vast facts.

Henri Poincaré, mathematician and physicist

Acknowledgments

First and foremost I'd like to thank the physicists, engineers, computing professionals, technicians, and other support staff who had the vision and determination to make the Large Hadron Collider and its associated detectors a reality. The LHC is one of the most complex scientific endeavors ever attempted, and I have the greatest respect for a group of people who can make it all work. As the scientific results start coming in, and certain people become known as the "voice of the LHC," we should never forget the teams that designed and built this equipment. Without them, those voices would be forever mute.

I would like to thank Dan Claes for contributing several hand-drawn figures for the text. He has helped me out in the past and I am very grateful, as if I had included my versions of these figures, well, it wouldn't have been pretty. I'd also like to thank Barry Panas and Jeffery Mitchell for various computer-generated figures.

I'd like to thank Leon Lederman for his gracious contribution of the foreword. Leon is one of the greatest living particle physicists, with more than one discovery that would have nominated him to the Nobel club. He is also a tireless cheerleader for basic research and spends more time in retirement crisscrossing the country, speaking with the public and policy makers alike than most people do at the height of their careers. The Energizer Bunny's got nothing on Leon.

I am deeply indebted to my test readers, without whom the text would have been vastly less readable. Linda Allewalt, Drew Alton, Lee Blakley, Rebecca Messer, Frank Norton, Chuck Osborne, Mandy Rominsky, and Michael Walsh all made invaluable suggestions as to language, scope, depth and breadth.

I also asked several colleagues to check that I had not typed in a wrong number when describing all the equipment. This is very easy to do, as the as-built numbers of a complex technical project such as the LHC and its associated detectors are often somewhat different than the formal design documents. Marzio Nessi checked the ATLAS section, while David Barney checked the CMS description. Yves Schutz and Roger Forty looked over the ALICE and LHCb sections respectively, while Michael Koratzinos vetted the accelerator section. In addition, I'd like to thank James Gilles for helping to identify these experts, each with a talent for public communication and a willingness to help out.

I should like to thank Tim Tait for doing the theoretical fact checking for yet another book. As always, his careful attention to detail was very helpful in ensuring that the most important aspects of the various theories I discussed were mentioned. I remain in his debt.

Of course, it is no doubt true there remain some errors in the text, no matter how valiantly these people worked to find them. These remaining errors are solely the responsibility of Fred Titcomb, who by virtue of his irresistible and evil mind rays, forced me to keep in a few mistakes. Between you and me, Fred is unaware I am writing this book, but I've known him for over 35 years and he was a convenient scapegoat back in kindergarten. Since I assigned him responsibility for errors in my last book, it would be rude for me to not keep up the tradition and not blame him here as well. Sorry Fred!

I am grateful to Bruce Schumm, who made some important introductions.

I absolutely must thank the staff at the Johns Hopkins University Press, starting with the editor in chief, Trevor Lipscombe. At my request, he pushed through the manuscript review process in what must be record time to allow the book to come out coincident with the turn on of the LHC. I should also like to thank the initial anonymous reviewer, who turned around the book proposal in just a couple of days. Michele Callaghan did a wonderful job in editing the original manuscript, polishing off the many rough edges. I should also like to thank the design and production and advertising staffs at Johns Hopkins and the typesetter for their roles in making this book a reality.

And finally, I must thank my family for putting up with my absences during this process and especially my wife for reading the very first draft and making important suggestions that really shaped the tone of the entire manuscript. The book is much clearer because of her input.

The Quantum Frontier

Prologue

Deep under the border between France and Switzerland, nestled between the primeval Jura Mountains to the north and the relatively youthful Alps to the south, a colossus stirs. When this giant fully awakens, it promises to reveal to mankind secrets long since lost to dim prehistory. The Earth has revealed ancient giants before. The nearby Jura Mountains lent their name to a period when Earth was stalked by beasts once long-forgotten: *Brachiosaurus, Stegosaurus,* and *Allosaurus.* But these denizens of the Jurassic era shook the Earth a mere 150 to 200 million years ago. The new awakening giant promises to teach us of a much earlier time, nearly 14 billion years ago. Indeed, it will tell us tales of the moment of creation itself. The giant stirring under the Swiss midlands is not a mythological beast but rather a scientific marvel, one of the wonders of the modern world. This book tells its story.

The CERN (the French acronym for European Nuclear Research Council) laboratory is one of the world's preeminent research institutions. Located just outside Geneva, Switzerland, it hosts physicists from all over the world who are working toward a common grand goal—unlocking the secrets of the universe. The centerpiece of CERN's research program is the world's largest and highest energy particle accelerator, designed to accelerate protons to nearly the speed of light and collide them in a controlled way. It began operations in 2008, with its full capacity coming online in 2009.

This accelerator has a name: The Large Hadron Collider, or LHC. Some two decades in the making, the goal of the LHC is to shed light on mysteries that so perplex those of us who think about what the universe is made up of and its origins: Why is the universe the way it is? How did we get here? Just what are the laws that govern the mass and the energy of the universe? Questions like these and many others are what drive physicists like me to dedicate our lives to seeking knowledge. These questions must have answers, which can be found if only we study them in the right way.

In this book, I hope to address these questions, and perhaps others, in five chapters. The first is a brief introduction into our current understanding of the

1

universe and the particles that make it up. This understanding, while impressive for both its breadth and depth, is far from complete. The second chapter describes a handful of the most important questions that the LHC is intended to answer and, perhaps more critically, just *why* these questions are considered important. The third and fourth chapters are geared toward those interested in truly understanding how we intend to use this marvelous scientific instrument to solve the mysteries, with the third focusing on the accelerator itself and the fourth describing the four big particle detectors being built for the task. The fifth chapter will look at the broader physics frontier. While the LHC will no doubt be the premier facility in the world for the next 15 or 20 years, my colleagues and I are already looking toward the future. In this final chapter, I will describe the expected playing field after the LHC has told us what it can.

Before I begin to address these questions, I want to dispense with a misconception that periodically rumbles across the Internet and through the media. Some people fear that when the LHC commences operations, it will endanger the Earth. There is, however, precisely zero risk.

Some worry that the LHC might create microscopic black holes, cousins of the monster black holes created in the death throes of massive stars. Stellar black holes have a gravity field so strong they would suck all nearby matter into them, not letting even light escape. If the LHC's higher energy might actually manufacture micro black holes, and from knowing how their stellar brethren work, people have suggested that microscopic black holes might swallow nearby matter in a runaway reaction that would devour the Earth. And, as my son succinctly put it, "Dude, that would *so* not be good."

Other people worry about other perceived threats. Some fret that the LHC will forge a kind of matter called strangelets, which would radically alter the Earth's matter. Others have brought up the possibility of creating a vacuum bubble. Their fear is that the universe is itself unstable and the LHC might trigger the cosmos to fall into a more stable state, in which the laws of nature might be quite different and in which life is no longer possible. Yet another danger claimed is that the LHC might make magnetic monopoles, which some theories claim would make the center of atoms unstable and the Earth and all the people on it would essentially evaporate. There have been many seemingly worrisome ideas put forth that suggest the only logical thing to do is to be safe and not turn on the LHC at all; better safe than sorry and all that. However, each of these worries has one thing in common.

They are all totally unfounded.

It is *impossible* that any of these scenarios are true. Even more comforting, we can be assured that there are no other Earth-destroying dangers posed by the LHC, even ones we have not considered. This is an important point. I could describe the particular reasons why black holes are not a problem and mention things like Hawking radiation and so forth. But even if you accepted my explanation on why black holes are not an issue, a skeptical reader might not be reassured, since the real danger might be posed by strangelets, monopoles, or left-handed floptwiddles. To understand just how safe the LHC is, you need to hear an argument that works *no matter what the potential danger might be.* Luckily there is a persuasive argument. We know we are safe because you are reading this book. Let me explain.

To understand properly why there is no danger, consider two important facts. First, the LHC will indeed collide beams of particles with unprecedented energy and intensity. However, although scientists talk about beams of protons, every collision in the LHC will be between exactly *two* protons, one from each beam. While the intensity of the beams make it more likely that two protons will collide with high energy in any particular second, there is essentially zero possibility that any collision will involve more than two.

The second fact is that the Earth is constantly being bombarded by cosmic rays from outer space and has been since its formation about four and a half billion years ago. Cosmic rays from outer space are most often protons that have been accelerated to very high energy by mechanisms we don't need to understand here. What we do need to know is that the energy of these cosmic protons can be as high as *and even exceed* those in the LHC. These cosmic rays hit the Earth's atmosphere and experience exactly the same sort of interaction that they will in the LHC, with a proton from an atom in the atmosphere of the Earth hitting a high energy proton from space.

In the eons since the Earth was formed, Nature has repeatedly pounded the Earth with cosmic rays, generating more collisions than the LHC would produce in many millions of years. That's *millions* of years. Indeed the cosmic rays are not limited to hitting the Earth. The universe as a whole generates in a single second 10 million million times as many high energy collisions as the LHC will over the next decade. And yet we're still here. If there were any danger, we wouldn't. No matter what happens in the LHC, whether micro black holes, strangelets, or some other dangerous-sounding phenomena exist or not, Mother Nature herself has conducted this experiment millions of times already. So sleep well at night and look forward with me to the bounty of discoveries that the LHC is sure to uncover.

But, for now, we begin our journey to the quantum frontier.

1

What We Know

The Standard Model

Science is a way of thinking much more than it is a body of knowledge.

Carl Sagan

We humans know a lot about the world in which we live. The origins of this quest for knowledge predate writing, as early man's very survival depended on an intimate knowledge of the natural world of seasons and plants, of tools and fire. Sheer pragmatism required that humans be keen observers. Almost certainly, there were early thinkers who wondered about deeper mysteries: those who wondered Why? as well as What? and How? We will never know just how deep ran the thoughts of these early scientists; however, we do know for certain that by 2,500 years ago, people were asking thoroughly modern questions.

On their craggy peninsula in the Aegean Sea, the early Greek philosophers debated long and hard about whether the natural state of matter was resting or moving and whether there existed a smallest particle of matter. Just as important, they recorded their thoughts so that others, separated by both space and time, could appreciate and build on their ideas and debates. In the recording, they tacitly laid claim to the origins of fundamental science.

Much has been written of these long-dead thinkers, but this book is not concerned with their specific thoughts. After all, their ideas were only generally correct and wrong in many specifics. However, we *are* concerned with their intellectual legacy.

Although the early Greeks may be credited with the start of the journey, the picture has been clarified in the intervening centuries. Our mastery of the natural world includes curing deadly diseases, learning to fly, and taking the first steps toward recreating the hot, all-consuming nuclear flame that fuels the sun.

In 1803, the British poet William Blake wrote "The Auguries of Innocence," which began

> To see a world in a grain of sand
> And a heaven in a wild flower,
> Hold infinity in the palm of your hand
> And eternity in an hour.

To see the world in a grain of sand is surely a metaphor, but it is not without an element of truth. By considering a single grain of sand and attempting to understand all of its fundamental pieces, one can learn a great deal about the laws that govern the greater universe. For instance, is there a smallest bit of sand? Under a microscope, sand looks a lot like a very small rock. If we crush the grain of sand, we are left with what appears to be even smaller rocks. If we crush those, do we have an infinite chain of ever-smaller rocks?

Asking this question for all the disparate substance of the world—rocks, water, air, food, and so on—led scientists to realize that all the matter of the universe could be created by combining different amounts of a little more than one hundred substances. We call these primordial substances *elements,* and some of their names are likely familiar from chemistry class, such as hydrogen, oxygen, and carbon. Combine hydrogen and oxygen, and you get water. Combine sodium and chlorine, and you get salt. In fact, if you mix the right elements in just the right way, you can make anything.

So one might ask whether these elements could be subdivided into individual units, that is to say, Is there a smallest unit of oxygen? And, indeed, it turns out to be true, with each element having a smallest piece. We call these smallest pieces *atoms* and have determined that the atoms of each element are distinct. If you want to have a basic mental picture of elements and atoms, think of an old-style toy store that specializes in selling marbles. One bin contains yellow marbles, while another has big red marbles and in yet another there are tiny green ones. So, each bin contains marbles of a distinct size and color. All the marbles within each bin are identical, and no two bins have marbles identical to those in any other bin.

So too it is with elements and atoms. All of the atoms of a given element are identical, and the atoms of different elements are distinct. And anything on Earth can be made by arranging the right combination of atoms in the right configuration. While the details of how you do the mixing are quite complex, one can learn a lot of chemistry just by using this simple analogy of marbles. Figure 1.1 lists the elements we've identified thus far in a chart known as the

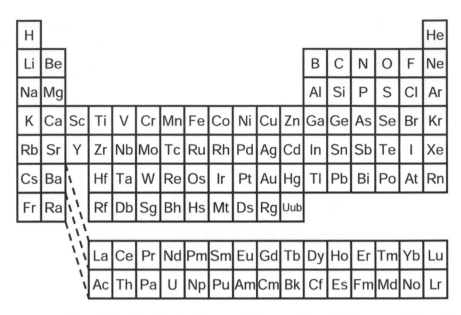

Figure 1.1. The periodic table, showing the currently discovered chemical elements. All observed matter in the universe can be constructed by combinations of these hundred or so elements.

Periodic Table of the Elements, or just the periodic table for short. Each block denotes a particular element. Elements that react similarly when combined with other elements are grouped together in columns.

Although what I've told you about atoms is true in important ways, in the early 1900s physicists came to realize that atoms could themselves be broken down. By 1932, physicists discovered that all atoms could be assembled through the right mix of three even smaller particles called *protons, neutrons,* and *electrons.* All protons were identical and so were all neutrons and electrons. On the face of it, this was a spectacular improvement in our quest for simplicity. The discovery that with about a hundred different kinds of atoms one could make anything in the world was an astounding simplification. But we now knew that the elements themselves could be made from the right combinations of these three simpler ingredients. For example, a hydrogen atom could be made from one proton and one electron, while helium atoms required two protons, two neutrons, and two electrons. The patterns for atoms of other elements were eventually determined.

As it became clear that atoms could be constructed of smaller particles, there naturally was interest in trying to figure out how the particles were arranged inside the atom. For instance, were the protons, neutrons, and electrons all clumped together in a tapioca-like mass? Or perhaps they were lined up like beads on a string. Logic really couldn't guide us to decide what an atom looked like. For that we needed experiments.

It was Ernest Rutherford, working at the turn of the twentieth century, who figured out the rough structure of the atom. He found that the atom is somewhat like a little solar system. From his and others' work, it was shown that each atom has equal numbers of electrons and protons. The protons are all clumped together with the neutrons in a tiny ball that is called the nucleus of the atom. The electrons swirl around the nucleus at a relatively great distance. Following the solar system analogy, the nucleus is equivalent to the sun and the electrons are more like the planets. The protons were found to have a positive electrical charge and the electrons had precisely the same amount of charge, but negative. Exactly why this should be so is not known even today. The neutrons were electrically neutral. Each atom had equal numbers of electrons and protons. The number of neutrons doesn't follow such simple rules, but, with the exception of hydrogen, the number of neutrons in an atom is similar to the number of protons but usually a bit higher.

After the basics of the atom were discovered, scientists learned other facts about its components. Even though the protons and electrons have equal electrical charge (although opposite in sign), they have radically different mass. The proton has about two thousand times more mass than does the electron. The neutron's mass is a smidge larger than the proton's mass. This disparity in the masses of the atom's components means that something like 99.95% of the mass of an atom is in the nucleus.

Protons and neutrons inhabit the nucleus of the atom, with the electrons swirling around at a relatively large distance, but this doesn't give us an accurate idea of the size of an atom. Atoms are really, really tiny. If you were to line up atoms "edge to edge," it would take 10 million to make up a single millimeter or 250 million to make up a single inch. Even after one realizes just how small the atom is, not even that truly gives the full picture. The atom consists of mostly empty space, with the diameter of the nucleus of the atom being about ten thousand times smaller than the atom itself.

One can perhaps get an idea of just how mind-bogglingly empty an atom is by analogy. Consider a carbon atom, one of the building blocks of life. A carbon atom consists of six protons and six neutrons in the nucleus, with six electrons swirling in a sphere, far from the nucleus. Let's imagine we blew up each proton or neutron to be a sphere the size of a printed "o" on this page. We could think of the nucleus as six of these red spheres (the protons) and six blue spheres (the neutrons) all clumped together. Let's further put this analogous nucleus at the 50-yard line of Soldier Field, home of the Chicago Bears football team. If we did this, the rest of the atom would consist of six electrons, each much smaller

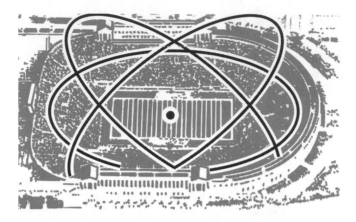

Figure 1.2. If protons and neutrons were blown up to the size of the letter "o" on this page, a single atom would fill a football stadium and yet most of this space would be empty. The relative size of the nucleus and atom are not drawn to scale. Courtesy Dan Claes.

than a printed period on this page, swirling like frenzied bees in a sphere the size of the football stadium. The atom is almost entirely empty space (Figure 1.2). Even so, these tiny, empty atoms of a hundred different elements, each consisting of only protons, neutrons, and electrons, form the building blocks of the entire universe.

Quarks

You'd think that scientists would celebrate the realization that with three tiny particles, they could explain the universe—and that they'd then leave well enough alone. But we physicists are a curious lot, and the scientists of the time kept poking at the question. In the 1940s and 1950s, physicists studied the data coming from their new toys, such as the "atom smashers," and from cosmic rays, which seemed to be raining down on Earth from space itself. They discovered particles in their data that did not fit neatly into the "proton, neutron, electron, or atom" classification scheme. In fact, they found nearly a hundred different particles that seemed to have similarities with the primordial protons, neutrons, and electrons. These particles were given names: pions, kaons, lambdas, and Vs. Scientists scratched their heads.

The scratching went on for quite a few years until 1964, when a very clever proposal was made. Maybe the primordial protons and neutrons weren't so fundamental after all. Perhaps they themselves were made of even smaller objects. These objects have come to be called quarks (pronounced "kworks"), after an inconsequential line from James Joyce's *Finnegans Wake* ("Three quarks for Muster

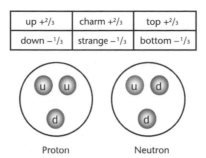

| up +^{2}/$_3$ | charm +^{2}/$_3$ | top +^{2}/$_3$ |
| down –^{1}/$_3$ | strange –^{1}/$_3$ | bottom –^{1}/$_3$ |

Proton   Neutron

Figure 1.3. The six quarks (*top*), with their fanciful names. The fraction indicates the charge held by that quark, where +1 is the charge of a proton. Protons and neutrons (*bottom*) are made by a suitable combination of up and down quarks.

Mark!"). Unlike earlier choices for the names of fundamental particles (both the words "atom" and "proton" have Greek antecedents: *atomos,* meaning "not able to be cut," and *protos,* meaning "first"), the word "quark" has no such academic inspiration and fits well with modern physics' tradition of whimsical names.

Originally only three quarks were proposed, but we now know of six. Their names are up, down, charm, strange, top, and bottom. These names don't really have any deeper meaning. Of all the quarks, two are by far the most prevalent: the up and down quarks. These two make up the proton (consisting of two ups and one down) and neutron (one up and two downs). The others are necessary to fully explain the plethora of particles discovered in particle accelerators (the pions, kaons, lambdas, and Vs listed above, as well as many others). Figure 1.3 lists the six quarks and shows how they make up the proton and the neutron.

The first three quarks proposed were the up, down, and strange quarks. The names "up" and "down" come from an older theory of the nucleus in which the protons and neutrons were treated as essentially the same thing. "Up" and "down" had a technical meaning but the words can be thought of as being similar to the two sides of a coin. The language of this older theory was carried over to the quarks. The name "strange" also is a historical holdover. Some of the particles discovered in the early accelerator and cosmic ray experiments acted oddly and people said, "Huh! That's strange." It turned out that the unusual behavior was related to the fact that they contained a strange quark within them, so the name migrated from the strange particles to the quark.

So it's a bit tricky to say when the up, down, and strange quarks were discovered, as scientists saw them in the first six or so decades of the 1900s. However, it was only in 1964 that they were recognized for what they were. The up quark has an electrical charge two-thirds that of a proton (+⅔), while both the down and strange quarks have only one-third the charge of the proton but with the opposite sign ($-\frac{1}{3}$). It seemed odd to have two quarks with $-\frac{1}{3}$ charge and only one with $+\frac{2}{3}$ charge, but that was how the theory was initially formulated.

The charm quark supposedly got its name because somebody said, "Wouldn't it be charming if there were a fourth quark, this one with $+\frac{2}{3}$ charge like the up quark?" It's hard to tell whether this is true or merely physics folklore, but the charm quark was simultaneously discovered in 1974 by two experiments, each based on one of America's coasts, at the Brookhaven National Laboratory on Long Island in New York state and the Stanford Linear Accelerator Laboratory in California. The bottom quark was discovered in 1977 at Fermi National Accelerator Laboratory (Fermilab) in Illinois, as was the top quark in 1995. I was one of the discoverers of the top quark as part of two competing teams of physicists, each comprising some five hundred scientists. The names top and bottom have no real meaning, although for a while "truth" and "beauty" competed for the honor of names for the two heaviest quarks. The use of these two alternative terms has declined over the past decade and is now pretty rare. That's kind of a shame, as I liked to tell people who came to my public lectures that I was "searching for truth."

With the introduction of quarks, we are approaching one boundary of the current frontier of knowledge. Thus it is important to pause to learn something of the nature of quarks. As best as we currently know, quarks are one class of fundamental particles. There are other types, and we'll discuss them shortly. In a physics context, "fundamental" means that to the best of our knowledge, quarks have no size and contain nothing smaller within them (i.e., they have no internal structure). Basically, in the journey into the heart of matter, we are made of molecules, which are in turn made of atoms. Atoms are made of protons, neutrons, and electrons, while protons and neutrons are made of quarks. But when we get to quarks, it's the end of the road. That's it. Quarks are as small as things get, or at least so goes current thinking. Figure 1.4 illustrates the various levels of the microworld for which we have some knowledge.

Now naturally, it may well be true that quarks actually are made of even smaller things. Such a possibility is just one of the exciting questions on which the Large Hadron Collider (from now on referred to as the LHC) might shed some light. We'll explore this possibility in the next chapter, but for the moment, let's concentrate on what we know about quarks.

Of the six types of quarks, two of them (the up and down quarks) are needed to make up protons and neutrons and, consequently, are stable, which means they don't decay. The other four quark types (charm, strange, top, and bottom) have very short lifetimes, existing for just a fraction of a second before decaying quickly into the more mundane up and down type quarks.

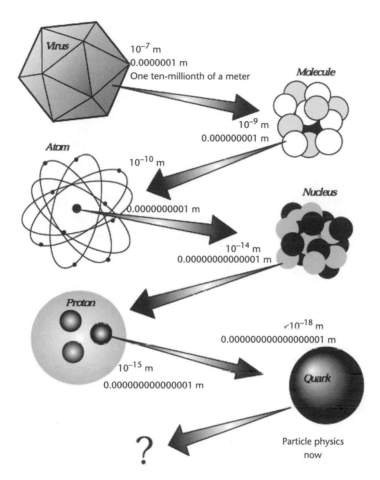

Figure 1.4. The study of nature involves looking at ever-smaller things to find the smallest and most fundamental constituents. Courtesy Fermilab.

Quarks have special rules governing how they can combine. As stated earlier, it takes three quarks to make a proton (two ups and a down) or a neutron (two downs and an up). We now know that to make any particle of the class that includes protons and neutrons requires exactly three quarks. There is another class of particles that can be made by combining one matter quark with one antimatter quark, but these are mentioned here only for completeness. Antimatter is a concept that will be described toward the end of this chapter.

To appreciate quarks, we have to peek ahead to the idea of forces. Although most people have at least a passing familiarity with gravity and electricity, far fewer people are aware that there are two other forces: the strong and the weak nuclear forces. These two forces, whose names we shorten to simply the strong and weak forces, have an appreciable effect only in the nucleus of an atom, with the strong force holding the nucleus together and the weak force governing some types of radioactive decay.

The strong force plays an important role in how quarks behave. Originally, the strong force was understood only as that which holds protons and neutrons together in the nucleus of the atom. There were earlier theories on how this force worked, but the picture was greatly simplified by the realization that quarks inhabit the protons and neutrons. It turns out that just as quarks have electrical charge and consequently feel the electrical force, they also have a new type of charge that governs the strong force. This strong force keeps the quarks in the protons and neutrons and holds the nucleus of an atom together.

Although this new type of charge is properly called the strong nuclear force charge, we colloquially call it "color." Color in this context has absolutely no relationship to the ordinary meaning of the word. We use the word "color" simply because of a convenient analogy. If you take red, blue, and green lights and simultaneously shine them on a wall, the resulting light will be white, which one might colloquially call no color at all. Similarly, individual quarks have color charge, but if you take them three at a time and put them in a proton, that proton has no net color charge. So we say that quarks can have three types of strong nuclear charge: red, blue, and green. Further, it is true that each proton and neutron always contain three quarks, each with a different color. It is not possible to have a proton with two or three red quarks, because protons have no net color, and only by combining red, blue, and green can one get white.

In Figure 1.5, we see that the color (strong charge) is unrelated to the quark type. For example, we can see that the down quark may have any color. To make a proton, all that is required is two up quarks and one down quark, each of which must randomly have one of the three strong force colors (red, green, or blue). Maybe this is easiest to see if we compare it to positive and negative numbers. For numbers, $(+1) + (-1) = 0$. For quarks, red + blue + green = 0 (or, equivalently, white).

In discussing color, we are led to another interesting feature of quarks. No quark has ever been directly observed. This doesn't mean that there is no evidence for quarks; indeed, the evidence for their existence is simply overwhelming. But it turns out to be impossible to pull a quark out of the atom and study it. Unlike a sandbox, from which you can pull out a single grain of sand to look at, quarks are locked firmly in their respective protons and neutrons. This fact is a consequence of how the strong force acts. The strong force is similar to a spring, in that as you stretch a spring, it gets harder and harder to stretch it more. Con-

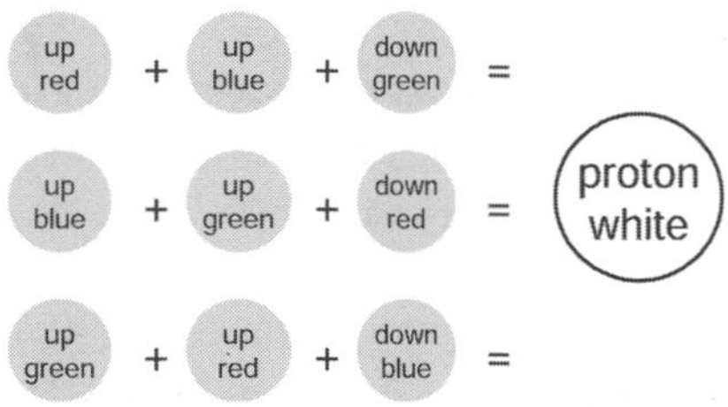

Figure 1.5. In quarks the colors red, blue, and green combine to make white. Similarly it takes three quarks, with three distinct strong-force charges to combine to make the strong-force neutral proton. The use of the word "color" for quark charges is purely metaphorical and has nothing to do with visible color.

trast this to the electric or magnetic forces, which get weaker as two charged particles are pulled apart. Think of two magnets, which get harder and harder to keep apart (or push together) the closer you bring them to one another. Conversely, when the magnets are far apart, they don't have any appreciable effect on one another. The springlike nature of the strong force has a very real effect on how quarks interact, but the most important feature is that quarks are generally stuck inside protons and neutrons. Technically, we say that quarks are "confined," which means that, under normal conditions, quarks cannot leave the proton or neutron that contains them.

The analogy between the strong force and a spring can be extended further. If you pull a spring or a rubber band hard enough, it will break. The strong force acts similarly. If you pull two quarks apart, the strong force resists more and more. But if you pull hard enough, the strong force "spring" will break. The distance at which the strong force spring breaks is about the size of the proton, which explains why the proton has the size it does. When the spring breaks, the quarks are then no longer connected and can move apart. Because of details beyond the scope of this book, these quarks are not "bare" quarks and cannot be seen like an electron that is knocked out of an atom. The idea is discussed in a little more detail in the text surrounding Figure 2.4. Briefly, in the "breaking" of the strong force spring, the energy originally stored in the spring creates more quarks and antimatter quarks. (This is a consequence of Einstein's oft-quoted but rarely understood equation: $E = mc^2$. Since the equation can literally be read as "energy equals matter," we see this as an example of energy converting directly to matter.) In the end, quarks always travel in pairs or triplets, safely ensconced in particles like protons.

The property of quarks that is most frequently mentioned is their mass, which spans a large range. The up and down quarks have a mass about 0.004

that of the proton, and the super-heavy top quark has a mass of 170 times that of a proton. We have only a hazy idea as to what gives the quarks their respective masses and, indeed, why they have any mass at all. The study of that particular question is perhaps the LHC's chief goal. In the next chapter, we will explore current thinking on this interesting question.

One thing that is very striking about quarks is that there seems to be a recurring pattern in their appearance. For instance, the up, charm, and top quarks all have the same electrical charge, as do the down, strange, and bottom quarks. Further, the up and down quarks are natural partners, in that they are the only quarks present in the stable proton and neutron. For this reason, as well as others, it is natural to group the quarks into three distinct pairs. We call these pairs *generations* and give each generation a number. The up and down quarks are generation I, charm and strange quarks form generation II, and top and bottom quarks form generation III. The reason for three similar groups of quarks is quite mysterious and is probably telling us something profound, if we only had the wits to understand it. Perhaps the LHC might teach us why this recurring pattern is present. We will get back to this question again in chapter 2. Table 1.1 summarizes what we know about quarks.

Leptons

We have identified protons, neutrons, and electrons as components of atoms and have identified quarks as components of protons and neutrons. So far, we've not discussed the role of quarks in the electron. That's because there are no quarks in electrons. In fact, like the quark, the electron is thought to be fundamental, which is to say that the electron contains no smaller particles within it. Electrons have electrical charge like quarks do, but they do not have color charge. Because of this they do not experience the "springy force" that quarks do and, consequently, each electron is not confined in the manner of quarks. This explains why they are not stuck in the nucleus but rather are free to orbit in the outskirts of the atom.

We said earlier that the universe can be built up by a proper mixture of up and down quarks and electrons. But we also know that there are two additional "carbon copies" (i.e., generations) of these quarks (e.g., the charm and strange and top and bottom quarks). Are there counterparts to the electron that might accompany these heavier quarks? Indeed there are. We have discovered two additional particles, called the *muon* and the *tau,* which have the same electrical charge and general characteristics as the electron but are heavier. Like the word "candy," which we use generically when we don't need to specify exactly what sugary food we're talking about, there is a word that allows us to refer to all electrons and electron counterparts. This word is "lepton," which stems from

Table 1.1 Names and characteristics of various subatomic particles

Matter Particles: Quarks						
Generation	**I**		**II**		**III**	
Name	Up	Down	Charm	Strange	Top	Bottom
Symbol	u	d	c	s	t	b
Charge[a]	$+2/3$	$-1/3$	$+2/3$	$-1/3$	$+2/3$	$-1/3$
Mass[b]	~0.003	~0.005	1.5	~0.1	170	4.5
Discovered[c]	1964	1964	1974	1964	1995[d]	1977
Lifetime[e]	∞	∞	10^{-12}	10^{-8}	10^{-24}	10^{-12}

Matter Particles: Leptons						
Generation	**I**		**II**		**III**	
Name	Electron	Electron neutrino	Muon	Muon neutrino	Tau	Tau neutrino
Symbol	e	ν_e	μ	ν_μ	τ	ν_τ
Charge[a]	-1	0	-1	0	-1	0
Mass[b]	~0.0005	~0	0.1	~0	1.8	~0
Discovered	1897	1956	1937	1962	1975	2000
Lifetime[e]	∞	∞	10^{-6}	∞	10^{-13}	∞

Force Causing Particles						
Force	**Strong**	**Electromagnetic**		**Weak**		**Gravity**
Name	gluon	photon	Z zero	W plus	W minus	graviton
Symbol	g	γ	Z^0	W^+	W^-	G
Charge[a]	0	0	0	$+1$	-1	0
Mass[b]	0	0	91	80	80	0
Range[f]	10^{-15}	infinite	10^{-18}			infinite
Strength[g]	1	0.01	0.00001			10^{-40}
Color	Yes	No	No			No
Discovered	1979	1905	1983			No
Particles affected	quarks	quarks, charged leptons	Quarks, charged leptons, neutrinos			all

[a]Electrical charge relative to a proton, which has a charge of +1.

[b]Mass in units, so the mass of a proton is 0.94.

[c]The up, down, and strange quarks had been observed (but not recognized) before 1964, which was the year the quark hypothesis was proposed.

[d]The author was one of the co-discoverers of the top quark.

[e]The lifetimes listed are in units of seconds and should be taken as representative only, as a quark's lifetime depends on its environment.

[f]The range is listed in units of meters.

[g]The strength of all forces is referenced to the strong force.

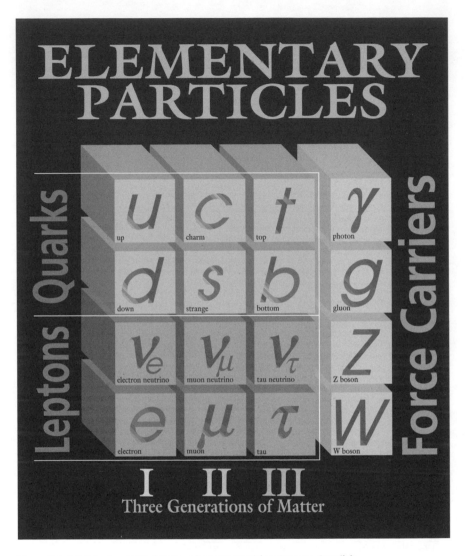

Figure 1.6. Full list of currently known subatomic particles. Courtesy Fermilab.

the Greek *leptos* for light. We generally refer to electrons, muons, and taus as charged leptons to remind us that these particles carry an electrical charge. Like much of physics, Greek letters are used to symbolize these objects. The symbol for the muon is μ (the Greek letter mu), while the symbol for the tau is τ (for the Greek letter tau). Table 1.1 and Figure 1.6 show how these charged leptons fit in to the scheme of subatomic particles. Some items in the table will be explained further below.

Although the electron (an electrically charged lepton) is a familiar particle, there also is a class of leptons that isn't so familiar. In the early 1900s, the study of radioactivity was all the rage. But there was a type of radioactivity that perplexed physicists. Radioactivity is the decay, or transmutation, of the nucleus of the atom of one element into the nucleus of another element. The confusion stemmed from when physicists looked at the energy involved in the process of decay, they found that the energy after the decay seemed to be lower than before. This fact flabbergasted physicists, as it was a fundamental tenet of physics at the time (and still is) that energy cannot be created or destroyed. Clearly something was awry.

The conundrum was solved in 1930 when Wolfgang Pauli realized that the radioactivity mystery could be explained if in the process of radioactive decay a particle was emitted that both had a very tiny mass and was electrically neutral. A name was proposed for the particle, *neutrino,* from the Italian for "little neutral one," because it was a neutral lepton. (Actually the name came from the Italian scientist Enrico Fermi, not the Austrian Pauli. Pauli's term "neutron" came to mean something else.) The neutrino was first experimentally observed in 1956. The symbol for a neutrino is ν, the Greek letter nu.

When Pauli proposed the neutrino, he didn't fully appreciate just how peculiar a particle it was. The reason the energy budget didn't add up in these peculiar radioactivity experiments was because the neutrino was carrying away some of the energy. Later experiments showed that neutrinos can pass through lots of matter without being detected. Although the penetrating power of neutrinos depends somewhat on their energy, neutrinos of the energies typically seen in radioactive decay could pass through five light-years of solid lead with just a 50% probability of being detected. Five light-years equals more than 48 million kilometers, or 30 million miles. So it's not at all surprising that the physicists doing those early radioactive decay experiments were unable to see the neutrino and were therefore confused.

Pauli spoke of only a single kind of neutrino, but a 1962 experiment showed that there is more than one kind of neutrino, with an electron-type and a muon-type clearly identified. Naturally, physicists wondered if there was a tau-type as well, a hypothesis confirmed in 2000. To distinguish the three types of neutrinos, we write them with a subscript (see Figure 1.6 for examples).

With the realization that there are three distinct types of neutrinos, each with an affinity for a particular charged lepton, our catalogue of the known types of matter particles is complete. Ordinary matter is made exclusively of up and down quarks, plus electrons and electron-type neutrinos. Why there should be two carbon copies (charm, strange, muon, and muon-type neutrino) and (top, bottom, tau, and tau-type neutrino) is not understood, but these

12 particles (six quarks, three charged leptons, and three neutral leptons) is the entire list of matter particles that we've discovered thus far.

Forces

We've listed the particles of which we're aware, but we've entirely neglected a crucial part of the story. After all, *something* keeps the planets circling the sun, electrons surrounding the nucleus of atoms, and the protons and neutrons firmly ensconced in their safe nuclear cocoon. These phenomena are governed by an idea called a *force.*

Forces can be simply defined as that which governs the motion of a particle. The force can attract or repel. Forces can even govern phenomena like radioactivity, which is kind of weird given the normal meaning of the word. In fact, we should use the word "interaction" instead of forces, so as to cover the radioactivity case. But the word "force" is so ingrained that we'll stick with it.

Physicists currently know of four forces (Figure 1.7). The most familiar of the forces is gravity, which keeps us solidly planted on Earth and governs the motions of the heavens. Ironically, this familiar phenomenon has most jealously guarded its secrets and remains the most mysterious force in the subatomic realm.

The second most familiar force is electromagnetism, which explains electricity of course but also magnetism, light, and all of chemistry. The electromagnetic force is much, much stronger than gravity and can cause both attraction and repulsion between two objects, while gravity is only attractive.

The other two forces are much less familiar. The strong force is responsible for holding the nucleus of the atom together, while the weak nuclear force is responsible for some kinds of radioactivity. As their names suggest, they have wildly different strengths.

Two important properties that distinguish the various forces are their ranges and relative strengths. Both gravity and electromagnetism have infinite range. In principle, every atom feels the effects of gravity from every other atom in the universe. In contrast, the strong and weak forces are only relevant over a very small distance and become essentially zero when the distances under consideration become larger than a proton.

With such different behaviors, there is no single number that characterizes the forces' relative strengths. After all, two quarks separated by a distance just larger than the nucleus of an atom would feel no effect from the strong or weak force but would feel the effects of both gravity and electromagnetism. But once we get two particles close enough that all four forces come into play, we can compare their strengths. In doing so, we find that these strengths span an enormous range.

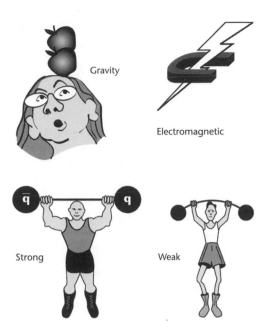

Figure 1.7. The four forces have distinct characteristic strengths and ranges over which they work. Courtesy the Particle Data Group, Lawrence Berkeley Laboratory.

For instance, if we take the strong force to be the standard against which we compare the other three, we find the second strongest force, electromagnetism, is about a hundred times weaker. The third strongest force, the weak force, is about a hundred thousand times weaker than the strong force. The most familiar of the forces, gravity, is weak enough to be in a class of its own, about 10^{-40} times smaller than the strong force. For those readers whose math is a bit rusty, remember that 10^{-2} is the same as 0.01. Thus 10^{-40} is a zero, followed by a decimal point, then 39 zeros and a one. That's small! Gravity is so weak that we've never been able to see any effect caused by gravity in modern particle physics experiments. Consequently, a quantum theory of gravity has eluded us. We simply don't know how gravity works in the realm of the ultrasmall. Further, the relative weakness of gravity is very troubling to physicists, and working out the reason for this weakness is something to which it is hoped the LHC might contribute.

People have a feel for how gravity works, but at the subatomic level, forces reveal a funny behavior. For "big" sizes, say about the size of a molecule, gravity is simply everywhere. Wherever you walk, gravity always pulls you downward, and there is no place where there is no gravity. In the quantum realm, things act differently. It turns out that, in the same way that atoms are small bits of elements, there are smallest bits of force. Each force has a characteristic particle associated with it.

The idea that a force like gravity could consist of small particles is somewhat counterintuitive, so let's explore it. Consider wind. Wind blows in your face, keeps a kite in the air, or pushes an empty can down the road. Wind exerts a force and is therefore analogous to other forces, like gravity or electromagnetic force. And, like gravity, air is something that is everywhere.

In addition to the forces, we also know something about chemistry. We know that air consists of molecules of oxygen, nitrogen, carbon dioxide, and the like. Thus the wind in your face is actually caused by uncounted molecules hitting you. Similarly, all of the forces at the subatomic level are treated as consisting of little particles of force.

As with much of particle physics, the names of the force-carrying particles are silly or simply mysterious. The particle causing the strong force is called the *gluon,* because it "glues" the nucleus together. The *photon,* familiar as light, is the particle carrying electromagnetism. Both the gluon and the photon have zero mass, but this isn't true for the weak force. Indeed, there are three types of particles that cause the weak force: the electrically neutral Z^0 (simply called "the Z boson") and two particles with electrical charge, W^+ and W^-, which are pronounced "W plus" and "W minus" (showing that they have the electrical charge of a proton [+] or an electron [–], respectively). These three particles are very heavy, with each one carrying a mass in the range spanned by bromine and zirconium atoms, or nearly a hundred times heavier than a proton.

The fourth force, the quantumly mysterious gravity, is thought to be caused by a particle, too. This particle is called the *graviton.* The graviton has never been observed, and you should regard with suspicion any claim to its properties. However if it *does* exist, we are able to work out what some of its properties must be. For instance it must be electrically neutral and have zero mass. Some day the graviton might be observed, and there's a Nobel Prize in it for the clever person who manages it. However, given gravity's weak nature, this prize is not likely to be claimed any time soon. Table 1.1 lists the details of the force-causing particles.

While the table lists the known quarks, leptons, and force-causing particles and brings us to the very frontier of knowledge, there is one little wrinkle that has not yet been mentioned. Even though we think the handful of particles and forces we've mentioned thus far is all that's needed to describe our universe, it turns out that there is a duplicate for every particle listed. Our next frontier topic concerns a mirror image of our familiar matter. This mirror matter is called

Matter   Antimatter

Figure 1.8. A paper clip made of matter combined with a paper clip of antimatter would release energy comparable to that released in the atomic detonation at Hiroshima.

antimatter, and it is one of the phrases popular with science fiction buffs that is science and isn't fiction.

Antimatter

The simplest description of antimatter is that it is the opposite of matter. Take some antimatter, add an identical amount of matter, and they both disappear in a blinding release of energy. The amount of energy released is huge compared with the amount of matter and antimatter involved. To give you a sense of size, if you took a paper clip made of matter and let it touch a paper clip of antimatter, the energy release is about the same as the 1945 atomic explosion at Hiroshima (see Figure 1.8).

Antimatter was predicted in 1928 by Paul Dirac. Does it really exist? The answer is a most emphatic Yes! The antimatter electron (called the *positron*) was discovered in 1932. The antiproton was observed in 1955, while the antimatter neutron waited until 1956. Protons and neutrons are made of quarks, but their antimatter counterparts are made of antiquarks. For example, the antiproton consists of two antimatter up quarks and one antimatter down quark. Antimatter particles have the opposite electrical charge from their matter counterpart; for instance, the proton has an electrical charge of +1, the antimatter proton (the antiproton) has an electrical charge of –1.

We have now observed antimatter counterparts for every type of quark and lepton. The simple existence of antimatter is interesting, but antimatter presents to us a truly fascinating mystery. To appreciate this mystery requires that you know two facts. First, you need to know that when we make antimatter in our laboratories, it always comes with an identical amount of matter. Always. Make an antimatter up quark, and you must simultaneously make an up quark. We never make an antimatter particle without a corresponding matter particle.

The reason for this is a thing called a conservation law. When there was only energy, there was no matter and no antimatter. Conservation means to keep something unchanged, so when antimatter is created, an identical amount of matter needs to be created to "cancel it out." Both the matter and antimatter are "created from nothing" or, more accurately, created from pure energy.

The second fact that one must consider is perhaps obvious, but extremely mysterious. This fact is the simple observation that in everyday life, we just don't see antimatter anywhere. There's nothing in our understanding of antimatter that excludes antimatter stars, antimatter planets, or even antimatter people. As long as these things were kept isolated from matter, they should exist. And yet they don't. Nowhere in the universe, as deep as our telescopes can see, do we see any substantial chunk of antimatter.

So why is that? Nobody really knows. This doesn't mean that we know nothing about the subject, but rather that the experiments done to date have not told us the entire story. We expect the experiments of the LHC to shed light on the mystery, particularly the LHCb experiment (described in chapter 4).

With the introduction of the quarks, leptons, force-causing particles, and now antimatter, we have discussed everything we know about subatomic particles. If we take the particles from generation I and tosses in the force-causing particles, we can build everything we see in the universe from galaxies to ice cream. Toss in the particles from generations II and III, and we can explain the results of all experiments ever conducted in our huge particle accelerators, too. We call the theory that includes all these ideas the Standard Model of particle physics.

With such a broad set of phenomena that we understand as well as we do, scientists are justifiably proud. To be able to take a handful of different types of particles and paint the tapestry of the cosmos is not a small feat. However, one should not be left with the impression that such an accomplishment has not left profound mysteries. In fact, for all our achievements, there's still a lot to do. Having focused our efforts on describing what we know, in the next chapters we shift our attention to some things we don't know and how the Large Hadron Collider is expected to shed light on these mysteries.

2

What We Guess

Theories We Want to Test

We all agree that your theory is crazy. But is it crazy enough?

Niels Bohr

Before we proceed, I should warn you that everything included here is completely speculative. We've left the comforting confines of what we know and have leapt into the unknown. At the frontier of knowledge, there is never certainty. Indeed, what we find in our experiments at the LHC may be similar to what we discuss below, or it may be something entirely different. Keep this in mind as you read. But this chapter does give you a good idea of what physicists wonder about as the LHC goes into operation and some of the things we think that we might find.

Although we know a lot about our universe, no one would argue that we know it all. Let's very briefly recap what we do know and see what sorts of questions are raised.

The observed universe is composed of two types of particles: quarks and leptons. Quarks are affected by all of the four forces: strong, electromagnetism, weak, and gravity. Leptons are not affected by the strong force, and a subclass of electrically neutral leptons, the neutrinos, is not affected by the electromagnetic force. We also know that there appear to be three identical generations of particles, with each generation containing heavier copies of similar quarks and leptons.

We also know about the four forces and that they have very different strengths, with gravity being ten thousand, trillion, trillion, trillion (about 10^{40}) times smaller than the strong force. Some forces are attractive, while others are both attractive and repulsive. Each of the forces (except gravity) has been shown to be caused by the transfer of subatomic particles, called photons, gluons, and the W and Z bosons. These particles can be electrically charged or neutral and can have either zero mass or considerable mass.

Another interesting piece of the story of forces is historical. In the past, our understanding of the nature of the world was less advanced than it is now. People saw that things fell when you dropped them. They also saw that the sun rose and set, the moon had phases, and the seasons came and went. These phenomena seemed to be unrelated, until a young genius by the name of Isaac Newton showed that the cause of all of them was gravity. We could say that Newton "unified" the behavior of falling things and the motions of the heavens with a single principle that explained both phenomena.

Similarly, although people have been aware of static charge, lightning, magnetism, and light for millennia, it was only in the 1800s that they were shown to be a single thing, now called electromagnetism. More recently, in the 1960s, physicists were able to show that electromagnetism and the force governing some kinds of radioactive decay (the weak force) were actually the same thing. Particle physicists now speak of the "electroweak" force.

This historical interlude leads us to the following question. While we speak of four forces (strong, electromagnetism, weak, and gravity), or three if we use the term electroweak, is it possible that further study will reveal that these seemingly unrelated phenomena are really all the same thing?

With these thoughts in mind, let's ask some questions:

- Why do the forces have such disparate strengths and ranges?
- Do the known forces end up being different ways to observe a single principle? If so, at what energy and why?
- Why quarks and leptons? Why do some particles have mass and others don't? Ditto electric charge? Why are quarks the only particles that feel the strong force? Why are there three generations? Could there be other generations?
- We live in a universe with three spatial dimensions and one time dimension. Why? Could there be more? What would they look like and, if they exist, why haven't we seen them?
- Why is the universe made only of matter, when we make matter and antimatter in equal quantities in our experiments? Where did the antimatter go?

There are other questions on which the LHC is expected to be silent or to comment on only indirectly. We'll sketch some of them in chapter 5. But the LHC is designed to explore the questions listed above (and many, many more), as well as to accurately measure familiar phenomena at the higher energies that only the powerful collisions of the LHC can provide.

A book like this cannot possibly address all these questions. Thus we will restrict the discussion to a few major topics, outlined below.

- What is the origin of mass and why do some particles have mass, while others don't?
- Will all the forces be shown to actually be the same thing, and why is it that current experiments hint that the energy at which this unification of forces might occur is so high?
- Why are there generations, and do they signal that there is something smaller inside quarks and leptons?

Finally, there are two additional questions that will be discussed here but will be given less attention. The reduced attention doesn't mean that they are of lower importance (after all, I've skipped some very important questions) but rather indicates that the LHC is not the only facility addressing these particular questions. But as the two more specialized of the LHC's detectors relate to them, these two questions will be raised here. One effort is the intensive study of particles that include bottom quarks, which scientists hope will shed light on why we don't see antimatter in the universe. The second is the study of what happens when nuclei of atoms of the element lead are slammed together at high energy. These studies will investigate what happens when matter is heated enough to allow quarks to freely escape their proton and neutron cocoons. We hope it will explore what conditions might have been during a period of the early universe about which we are currently largely ignorant.

It's completely wrong-minded to say that "the LHC was built to discover X." That would mean that "X" is understood well enough to know that it's there and therefore to find it isn't really a discovery. No, the purpose of the LHC is to study the nature of matter under conditions that are seven times hotter and more energetic than ever before observed. We will see what we see. Interesting, fascinating, or disappointing, the universe will reveal some of her secrets, and the world will become slightly less mysterious.

Scientists could not have persuaded the world's funding agencies to support a multibillion-dollar endeavor if they didn't have a very good reason to expect that there *would* be valuable discoveries. Probably the most likely and anticipated discovery for the LHC's experiments is the explanation of why subatomic particles have mass. Rather counterintuitively, this is related to understanding how the electromagnetic and weak forces are one and the same.

The story of our understanding of the origins of mass has a complex history. It begins in the 1960s, when a bunch of young physicists were working to see if the electromagnetic and weak forces might be two sides of the same coin. When you get right down to it, this wasn't such an obvious thing to do. After all, the weak force's strength is about a thousand times smaller than the electromagnetic force's and, further, the two forces have very different characteristics. For instance, the electromagnetic force has an infinite range, while the weak force's range is very short and only felt over distances about a thousand times smaller than a proton. Also, for a particle to feel electromagnetism, the particle must have electric charge. Particles that feel the weak force can be electrically neutral (e.g., the neutrino).

Early in the 1960s, the weak force was not known to be governed by the exchange of particles in the same way that the electromagnetic force was governed by the exchange of a photon. However, by knowing the range over which the weak force is felt, physicists could calculate the mass of the weak force particle if it existed. The result was that the weak force particle had to have something like a hundred times the mass of a proton (which is considered huge in the particle realm even now and was almost unthinkable at the time). Given that the electromagnetic-carrying photon was known to be massless, this goal of unifying the weak and electromagnetic forces could well have been impossible.

So the physicists of the day did what physicists do. They made a simplifying assumption. Suppose that the mass of the weak-force carrying particle was zero like the photon. What then? Well, through an intellectual tour de force, it was accomplished; the electromagnetic force and an "almost correct" version of the weak force were shown to be governed by a single equation. This equation predicted four massless particles involved with the newly understood electroweak force.

The actual history of this triumph is beyond the scope of this book, but it can be found in the suggested reading. The story, like most big scientific discoveries, had many heroes, although too few villains to make a topnotch movie. These physicists made false starts and made brilliant insights, both successes and failures, and by 1970, the basic understanding was in place. Physicists then predicted the heavy particles (the W and Z bosons discussed in chapter 1) that are the source of the weak force. In 1983, these particles were observed for the first time, validating the theory. Everybody was happy.

However, you might ask, "How do you get from the four massless particles discussed two paragraphs ago to the four observed electroweak particles: photon, Z boson, and the positively and negatively charged W bosons, only one of which is massless?" Understanding this link is one of the primary goals of the LHC.

A Scotsman to the Rescue

In 1964, Peter Higgs, a Scottish physicist, followed a suggestion from Phillip Anderson and proposed that perhaps the universe was filled with a new kind of field. This field has come to be called the Higgs field. To get a feel for an energy field, think about the gravity here on Earth. Gravity is everywhere. It passes through everything. So too it is with the Higgs field. Then the question arises, "So what?" What does the Higgs field do, and why is it interesting? Further, how does the Higgs field solve the problem of the origin of particle mass?

To get an idea about how the Higgs field comes into play requires two crucial ideas. The first is the idea of an add-on or modifier. The idea is pretty simple. The world is a complex place, and physicists like simplicity. For instance, physicists always say that all objects fall at the same rate. Drop a marble and a bowling ball from the same height, and they'll hit the ground at the same time. This is an experiment you can do and, after a little practice in simultaneously releasing the two objects, you can see that it is true.

Yet my students don't really like the assertion that "all objects fall at the same rate." They correctly point out that if one drops a hammer and a feather, they fall at quite different rates. This observation makes them unreceptive to further learning. I even show them the video of an Apollo astronaut dropping a feather and a hammer on the moon, where the two objects do indeed fall at the same rate, to no avail. And yet this video illustrates the idea of the add-on. There is no air on the moon and there is on Earth. It is air friction that invalidates this simple statement about gravity.

Yet the statement isn't wrong. Gravity does cause all objects to fall at the same rate, as evidenced by the lunar video. It's just that gravity isn't the entire story. You need to include the effect of air friction to get a more accurate prediction of reality. Similarly, in the particle world, the equations that involve massless particles are also correct to a point. But it takes the Higgs field to account for the observed particle masses.

The second crucial idea is the idea of symmetry and how to break it. Symmetry is a mathematical term, describing equations. However, the idea is simpler and more universal than that, and we can understand it without using any math at all. Symmetry is when something looks unchanged after a change is made.

Figure 2.1 shows a circle and a square. The circle is the most symmetrical two-dimensional object; rotate it any way you want and it looks the same. The square is symmetrical, too, just less so. If you rotate the square by anything other than a multiple of 90°, you can see that it has been disturbed. However rotate the square by 90° (or 180° or 270° or, well, you get the picture), and you're back to a situation that is indistinguishable from where you started.

In the math sense, the equations are said to be symmetrical if you can swap the symbols around and end up with the same equation. So in the case of the electroweak equation, if you swapped the symbols denoting the various force-carrying particles, it doesn't matter, the equation is unchanged.

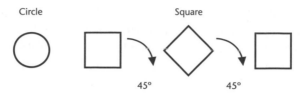

Figure 2.1. A circle can be rotated by any amount without looking different, while a square has very specific rotations after which the change is not evident.

To break symmetry is to do something that makes it obvious that a change has occurred. Suppose that you have a table with two chairs and two people sit in them, facing one another, as shown in Figure 2.2. As far as the two people are concerned, it doesn't matter who sits in what seat; the two people are always facing one another. But now put three seats at the table and use three people. Now if two people swap seats, everyone can tell. This is because the people involved can tell that the others have moved from their left-hand to their right-hand side. The addition of the chair has broken the symmetry. In the particle case, we say that the addition of the Higgs field has made it possible to identify which symbols denote the massless photon and which the massive Z^0 boson.

So let's get back to the Higgs field. The Higgs field is an add-on to the simpler theory, just like air friction is an add-on to gravity when describing how things fall. The basic idea is that different particles will interact differently with the Higgs field. Massive particles interact more with the Higgs field, while the massless photon is not affected by the Higgs field at all. In fact, particles are massive *because* they interact with the Higgs field. It is this interaction that gives them their mass. Please note that in the context of physics, the word "massive" connotes something completely different from its usual meaning. It indicates that we are looking at a particle with mass, as opposed to a massless particle such as a photon.

Think of our air friction analogy. A falcon can cut through the air with the greatest of ease, plunging swiftly from great heights to catch its prey. The air has little effect on the falcon's descent. Compare the falcon with a guy with a parachute. The parachute interacts a lot with the air. The air friction that stands in for our Higgs field gives the parachutist great mass and gives almost none to the falcon.

So thus far we've described what the Higgs idea means. The next and most important question is, "Why should we believe it is true?" And that's a good question, as the Higgs idea is unproven. Even Higgs' original 1964 paper submission was rejected, since it predicted nothing new. It was only after he added a

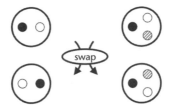

Figure 2.2. In the case of two diners (*left,* shown by different circles), if they swap seats, there is no apparent change, as they still see each other as sitting across the table. However, in the event of three diners (*right*), a seating swap is obvious as people now can tell that their companions have moved from their left to right. The symmetry is broken.

sentence to the end of the paper that it was accepted. This sentence noted that if his hypothetical field were true, then one experimental consequence followed. He predicted a new particle.

The correspondence between a particle and a field is not an obvious one, so although we mentioned this topic in chapter 1, let's take a brief detour to explore the idea again. Everyone is familiar with air; it is everywhere. It is gaseous and fluidic and permeates everything. You can't take a coffee cup and scoop out some air and leave a hole behind. You can think of air as a continuous field in which we live.

And yet, it's not. You also know that air is composed of molecules: there are "smallest bits of air." We can say without too much sloppiness that there are air particles. If we simultaneously hold in our head the idea of a pervasive fluidic air and air molecules, then it is easy to hold in our head the idea of a field and an associated particle. A more scientific, but relatively familiar, example is an electric field. If you rub a latex balloon on your shirt and run the balloon just over your arm, you'll feel your arm hairs affected by the electric field. This field fills the space between your arm and the balloon.

Even though the electric field is somewhat familiar, you need to remember that the electric field is composed of countless photons, just like air is made of individual air molecules. Similarly, if there is a Higgs field, there is a Higgs particle. This particle, if it exists, is called the Higgs boson.

The Higgs boson is predicted to have rather specific properties. It is electrically neutral. It has no size and no structure and yet it has mass. No structure means that we do not believe that there are any smaller particles within it. Earlier in the book I referred to such particles as fundamental. It is a scalar particle, which means it doesn't spin. My scientific colleagues will cringe a bit at that, since "spin" in the quantum realm is subtly different from the ordinary meaning of the word. But, for our purposes, we can forgo the distinction and simply say "Higgs bosons don't spin," which is in contrast to every other known subatomic particle. We would then say that the Higgs boson is an electrically neutral, massive, fundamental scalar.

If such a particle did exist, how would you see it? Like all short-lived particles, you would not see it directly but rather its longer-lived decay products. We are in a tricky situation in attempting to describe the Higgs boson. As stated earlier, it has not yet been observed. However, we can predict many things about it and its properties. For example, we can predict that when the Higgs boson decays, it will generally decay into two particles with opposite electric charges. Further, because the Higgs idea is integrally related to mass, the Higgs boson will generally decay into the heaviest pair of particles it can.

So, if we look at Table 1.1, we can find the list of known particles and identify those that have the most mass. They are—in descending order of mass, measured in billions of electron volts (GeV) in which each GeV is about the mass of a proton—top quarks (175 GeV), Z bosons (91 GeV), W bosons (80 GeV), and bottom quarks (4.5 GeV). Because you can't get something for nothing, the Higgs boson's mass must be at least twice the mass of the objects into which it decays. For instance, to make two 175 GeV top quarks, you need a minimum of 350 GeV to start.

The reader may wonder why the unit "electron volts" are used to measure the quantity of mass, instead of the more familiar pounds or kilograms. An electron volt is a unit of energy gained by an electron when it is accelerated by a one-volt electric field. Since Einstein showed that matter and energy are the same, we can freely interchange their units and use the term electron volt to describe mass. This is an incredibly convenient choice for particle physicists and allows us to easily understand the linkage between the strength of our accelerator and the mass of the particles that we can make using it.

Because scientists believe the Higgs boson prefers to interact with heavy particles, we believe it will decay into the heaviest particles that it can. For instance, if the mass of the Higgs boson is more than 350 GeV, it can decay into a top-antimatter/top-quark pair. If its mass is above 182 GeV, it can decay into pairs of Z^0 bosons. Above 160 GeV, the daughter particles (or particles created as a result of decay of a given "mother" particle) will be a $W^+ W^-$ pair, while between 9 and 160 GeV, the way to look for Higgs bosons is to try to find bottom-antimatter/bottom-quark pairs.

When one looks very carefully at the theoretical predictions, one sees that the situation is slightly more complex, as illustrated in Figure 2.3. The Higgs bosons definitely can decay in the ways described in the previous paragraph. But subtle physics effects also come into play that give an edge to W and Z bosons in the competition as to how the Higgs boson will likely decay. The net effect is

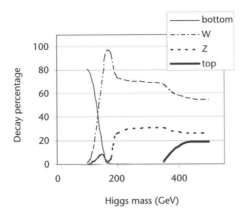

Figure 2.3. Predicted decay percentages of the Higgs boson. If the Higgs boson mass is below 135 GeV (billion electron volts), it preferentially decays into bottom-antibottom quarks. Above that threshold, pairs of W bosons are preferred. Above 190 GeV Higgs bosons are predicted to decay into pairs of Z bosons about 30% of the time. If the mass of the Higgs boson is larger than 350 GeV, it can also decay into a pair of top quarks.

that below a possible mass of the Higgs boson of 135 GeV, it will decay predominantly to bottom/antibottom quark pairs, while above that the preferred decay mode is into W boson pairs. However, the other decay modes we've listed will be possible, in addition to ones not mentioned here. LHC physicists will be looking for all of these possible types of pairs.

Given that bottom quarks, W and Z bosons, and top quarks are the most likely particles into which the Higgs boson will decay, let's explore a little bit about how these decay particles will themselves be observed in a detector. After all, we will never see the Higgs boson itself but only infer its existence from its daughter particles. The problem is that all of these daughter particles decay as well. Top quarks decay 100% of the time into bottom quarks and W bosons. So this gets us to the final point. Any LHC detector that wants to look for the Higgs boson predicted by current theory had better be able to measure W and Z bosons and bottom quarks well. So let's briefly examine how Z bosons, W bosons, and bottom quarks decay.

Because of the nature of the strong force, quarks don't like to be alone. If a quark is pulled away from other quarks, the strong force acts a bit like a glob of water thrown from a glass. First there is a slug of water, but surface tension pulls it apart into individual water droplets. Figure 2.4 illustrates this analogy.

Thus a single quark will turn into many particles, all traveling in the same direction. This stream of particles is called a "jet" and resembles the blast of pellets that come out of a shotgun. Just as a single shotgun cartridge turns into

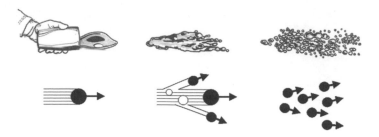

Figure 2.4. Just as a slug of water will turn into water droplets, a single quark will turn into many particles. The physical mechanism is quite different, but the essentials of the process are quite similar. Water drawings courtesy Dan Claes.

many pellets, all going in generally the same direction after they leave the barrel, a single quark turns into many particles all traveling in the same direction after they leave the vicinity of other quarks. Obviously, the physical mechanisms governing the two phenomena are very different, but the mental picture is very valuable.

All quarks turn into jets after a particle collision except for top quarks, which turn into bottom quarks and W bosons before there is time for a jet to form. But the daughter bottom quarks that result from the collision do form a jet.

The Z and W bosons can also decay into quarks. But these bosons have an option not available to quarks. The bosons can also decay into pairs of leptons, of which the electron is the most familiar. Figure 2.5 shows the various ways that the Z and W bosons can decay. From an experimenter's point of view, the most interesting decays are those involving electrons, muons, and neutrinos. These are interesting because they are very distinct and generally indicate that a W or Z boson was created in the collision.

So let's combine our information. Suppose the Higgs boson is very light, say 115 GeV (or about 115 times greater than a proton). We see in Figure 2.3 that the Higgs boson will most frequently decay into a bottom quark and an antimatter/bottom quark pair. Thus to see a light Higgs boson, you need a detector that can measure two jets very well, both coming from bottom quarks.

If the Higgs boson is much heavier than a proton, say 160 GeV, then the Higgs boson most likely decays into a $W^+ W^-$ pair. Since jets can come from quarks from W boson decay as well as more ordinary quarks just getting knocked out of the proton (and jets from ordinary quarks are much, much more common), experimenters are more interested in the decays involving the leptons. Say one W boson decays into an electron and neutrino, and the other decays into a muon and neutrino. Since both the Higgs and W bosons decay rapidly, what we as experimenters would see is an electron, a muon, and two neutrinos. Since

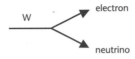

W decay modes	
Decay mode	Frequency
W→quark + antiquark	67%
W→electron + neutrino	11%
W→muon + neutrino	11%
W→tau + neutrino	11%

Z decay modes	
Decay mode	Frequency
Z→quark + antiquark	70%
Z→neutrino + antineutrino	20%
Z→electron + positron	3%
Z→muon + antimuon	3%
Z→tau + antitau	3%

Figure 2.5. *Top,* representative decay. The tables (*bottom*) show the frequency by which the W and Z bosons can decay. The missing 1% in the table at the right is a rounding-off error.

neutrinos don't interact very much with matter, they escape undetected. So we would see an electron, a muon, and missing energy.

Missing energy makes things a little tricky. Even though it is thought that Higgs bosons decay into W boson pairs about twice as often as they do into Z boson pairs, that Z bosons decay into pairs of charged leptons make that particle a very attractive way to look for Higgs bosons. For instance, if a Higgs boson decays into two Z bosons, and both Z bosons decay into electron/antimatter-electron pairs, the experimental signature of a Higgs boson would be two electrons and two antimatter electrons (positrons) in your detector. If one of the Z bosons decayed into a muon/antimatter-muon pair instead, then you'd see a pair of electrons and a pair of muons. The most important point is that when a Z boson decays, you see both decay particles, which you wouldn't for the W boson.

Obviously the experimenters at the LHC will look at all possible ways in which the Higgs boson can decay, keeping in mind the various ways in which its daughter particles may themselves decay. Studying all these possibilities will keep a small army of physicists busy for quite some time. There is a way the Higgs boson can decay that is attractive to physicists. Higgs bosons decay into a quark and antimatter quark pair. These two particles can touch each other, annihilate, and emit two photons. Sifting through events in which two photons are produced may well be the way in which the Higgs boson is discovered.

While I've told you how people intend to search for the Higgs boson, I've not told you where we're going to find it. That's because we don't know. As of the summer of 2008, the direct evidence for the Higgs boson is zero. That's right . . . zero.

So why do we believe the Higgs boson or something like it exists? Because the circumstantial evidence is high. The electroweak theory—recall this is a theory that describes the behavior of photons and the W and Z bosons—has been tested to exquisite precision. The Higgs boson is an integral part of the theory, even though it is only an add-on. Given the precision with which the theory has been tested, we have reason to believe the theory, including the Higgs boson. Of course, "reason to believe" is not good enough. You need proof. Taking a legal analogy, it's the difference in the level of proof needed to charge someone with a crime compared with the burden of proof needed to convict that person. All we *really* know is that the symmetry between the weak and electromagnetic force is broken. One (and the most popular) candidate is Peter Higgs' original idea. However, there are others, one of which I will mention briefly in a moment.

Even without ironclad certainty that Higgs' idea is right, we can still ask sensible questions. The following question may sound weird to someone who has taken high school or even college science classes, because in those classes uncertainty is not encouraged; science (and the teacher) knows all the answers. However, at the research frontier, uncertainty is rampant. In fact, much of a scientist's postcollege education is dedicated to learning how to handle uncertainty in a rigorous way. So the question that researchers ask of the Higgs idea is the following: "Suppose that the Higgs idea is correct. If so, what do existing measurements tell us about the Higgs boson's properties?"

We know that any possible Higgs boson mass that goes against the current electroweak theory can be ruled out. That means we can guess a Higgs boson mass, put it in our equations, and calculate something involving W or Z bosons. If that prediction turns out to disagree with our measurements, we know that was a bad guess. By using this kind of indirect logic, we know that the Higgs boson, if it exists at all, has a mass between about 50 and 175 GeV. That's a big range, but it's something. At least we can rule out zero, 500, and 1,000 GeV as possible masses for the Higgs boson.

However, we've not exhausted all ways in which we can limit the possible masses that the Higgs boson could have. Not only have we tried the "guess a Higgs boson mass and see its effects on other things" approach, but we have also looked directly for the Higgs boson. We have determined that if its mass is below 135 GeV or so, we just need to look for decaying events with two bottom quark jets.

Before the LHC was constructed, another particle accelerator inhabited the same tunnel. This accelerator was called LEP (for Large Electron Positron), and it supplied electron and positron beams to four superb detectors. Together, they looked for Higgs bosons and failed to find any. However, even though these detectors didn't find any Higgs bosons, they had the ability to recognize the bosons if their mass were below the precise number of 114.4 GeV.

So, as of the summer of 2008, we know that if the Higgs boson exists that it will have a mass in the range of 114 to 175 GeV. The most likely spot is about 125 GeV or so. And, we know if it exists at all, the LHC will find it.

But "if" is the operative word. The LHC is a discovery machine. There are no guarantees. Just like Columbus could not predict that he would find the New World—indeed he predicted he would find something else entirely—neither can scientists working with the LHC say with certainty what they will find.

In fact, the Higgs idea is not the only one that could explain electroweak symmetry breaking. A theory called technicolor explains electroweak symmetry breaking in an entirely different way, involving fermions (particles with a half-integer spin such as electrons and protons) instead of Higgs bosons. Recall that Higgs bosons are believed to be scalar, or having no spin. We'll come back to the concept of spin later in the chapter.

There are also theories that predict a Higgs-boson-like particle, except in these theories, the Higgs boson is itself composed of particles within this purported particle. One idea is that the Higgs boson consists of top quarks in a manner similar to how the proton contains up and down quarks. Until we make measurements that tell us how the universe actually acts, we cannot definitively rule out any of these ideas yet.

Another theory that can explain electroweak symmetry breaking is one involving a principle called supersymmetry. A search for supersymmetry is another big focus for the LHC, and one we will discuss extensively soon. Theories involving supersymmetry can predict electroweak symmetry breaking in ways that don't have that clumsy "add-on" flavor of the Higgs idea. Of course, supersymmetry predicts a whole slew of particles not yet observed, including more than one Higgs boson. This theory further suggests that these other Higgs bosons have an electric charge.

The bottom line is that electroweak symmetry breaking is not understood and, while the Higgs boson idea is the most popular of the ideas put forward to explain it, it is by no means a foregone conclusion that Professor Higgs' boson is the answer. Only through experiments can we know. Assuming that there are no unforeseen difficulties, this question should be resolved within five years of the startup of the LHC and, we hope, even more quickly.

Supersymmetry

Mathematical principles govern the universe and are embodied in the equations that physicists use to make their predictions. The most critical of these principles are called *symmetries*. There are many kinds of symmetries, some well

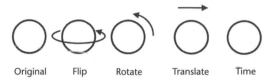

Original   Flip   Rotate   Translate   Time

Figure 2.6. Circles are the most symmetrical two-dimensional object. Do nearly anything to them, and you can't tell that something happened.

established and with which the reader is very familiar, and some rather nonin-tuitive and about which it is hoped that the LHC will say something profound. Before we can understand this particular LHC goal, we need to spend time defin-ing some basic ideas.

The first idea is that of symmetry itself, a concept that we touched on ear-lier. Symmetry is both a mathematical and a visual or artistic concept. Basically in both math and art, a symmetry is something you can change and nobody will know.

A circle is the most symmetrical two-dimensional object. (We pick two-dimensional because this page is two-dimensional. We could as easily use three-dimensional, but then the symmetrical shape would be a sphere.) In Figure 2.6, we see that when we start with a circle, we can flip, rotate, move, or look at it again tomorrow and not tell the difference. Technically, we say that the circle is "symmetrical" under all these possible changes. For the more mathematically inclined, we'd say that the shortest distance between the center and a point on the perimeter is unchanged under these operations. See the discussion sur-rounding Figure 2.1 for a figure that isn't quite as symmetrical, the square.

In physics, symmetry has a similar meaning, but it also has an added physi-cal significance. Imagine a bug at the bottom of a bowl. To crawl out, he'll have to do a lot of work. If he crawls out by going to the right, he'll have to do a cer-tain amount of work. If he tries to crawl out to the left, he has to do the same amount. Here's the important thing: If you switch the words "right" and "left" in the last two sentences, nothing would change. Similarly, if you moved the bowl across the room or moved it from the table to the floor or vice versa, the bug's predicament would be identical. And, assuming that he brought a snack and a little bug sleeping bag, so he'd be well fed and rested, his effort needed to get out would be the same tomorrow as well.

Getting beyond examples of bugs and bowls, physicists can write equations describing the standard model, which you recall is the theory that embodies our current understanding of the universe. In fact, the standard model equa-tion includes all possible symmetries save one. This additional symmetry deals with how particles spin at the quantum level. Before we finish introducing this interesting new symmetry, we need to go over a couple of things about spin and quantum mechanics.

We are accustomed to thinking about certain aspects of the world as being "quantized," or coming in discrete units. Electrons come with only one mass. It is impossible to have an electron with half the mass of the others. Similarly, electrons only have one charge. However, electrons have another feature that is not as quite as familiar. Every electron in the universe is spinning identically. This is counterintuitive, since we are used to objects being able to spin faster or slower and not having a single amount of spin that is allowed. In contrast, we're also accustomed to being able to change mass at will. A wheelbarrow of sand can have more or less weight, depending on whether we toss in that final shovelful. And yet each electron has an identical mass. So the disparity between the standard concept of spin and the quantum concept maybe isn't so hard to accept.

The numerical amount of spin isn't so important. (For the technically minded, the spin of the electron is $\frac{1}{2}\hbar$, where $\hbar$ stands for the Planck constant and is a tiny number.) We can ignore its numeric value and just recall that all spins are expressed in units of $\hbar$ and therefore simply drop the "$\hbar$", calling the spin of the electron $\frac{1}{2}$. Not only is it true that all electrons have a spin of $\frac{1}{2}$ but the same is also true of quarks. In contrast, the force-carrying particles—the W and Z bosons, the gluons, and the photons—all have a spin of 1, or twice that of the quark and electron. The Higgs boson, if it exists, has a spin of 0.

Further study has revealed that there are two fundamental classes of particles: the fermions, with half-integer spin (e.g., $\frac{1}{2}$, $\frac{3}{2}$, $\frac{5}{2}$, and so on), and bosons, with integer spin (e.g., 0, 1, 2, 3, and so on). You'd think that a little thing like a half unit of spin wouldn't make all that much difference, but it does. Fermions, which include quarks and electrons, are the rugged individualists of the particle world. No two fermions can be in the same space at the same time and with the same energy. This has huge consequences for chemistry, which, after all, is the study of the fermion electrons around atoms. For those who have taken chemistry, this is the source of the Pauli exclusion principle and explains a great deal about the structure of the periodic table.

Bosons, on the other hand, are gregarious. "The more the merrier" is their motto. Two bosons can be in the same place at the same time and with the same energy. No problem.

Getting back to symmetries, there remains a possible symmetry not yet observed. It is called supersymmetry, often denoted by the letters SUSY, and is predicated on the idea that you could exchange fermions and bosons everywhere in the equations (and in the universe) and nobody would notice the difference.

Well, this symmetry has not been officially added to quantum mechanics for the simple reason that it's absurd. If we replaced the observed fermionic electrons with a boson equivalent, all of chemistry would be radically different.

Nonetheless, mathematically at least, one can think of constructing a physics theory that includes supersymmetry and in which you can swap all fermions for bosons and vice versa. It is important to note that supersymmetry is not a theory. It is a principle. It's like "conservation of energy" for the scientifically minded or "greatest good for the greatest number" for the philosophers. There are many ways one might try to apply this guiding philosophical principle, with some adopting a "Mother Teresa" behavior to help the poor and others adopting a "Bill Gates" behavior and setting up a philanthropic foundation. The principle is "greatest good," while a theory would be the Mother Teresa or the Bill Gates individual implementation of the principle.

Similarly, in physics, many specific theoretical models incorporate the principle of supersymmetry. But these theoretical models are not supersymmetry, per se. Supersymmetry is much bigger than that.

In 1981 someone took the conventional standard model and added supersymmetric principles to it. This new model is called the minimal supersymmetric standard model, or MSSM. As its name suggests, MSSM is the usual standard model, with the absolute minimum number of changes necessary to incorporate supersymmetry. It would be easy to overcomplicate what was done but in essence scientists added terms to the standard model equation. The standard model has terms for the matter fermions (e.g., the quarks and leptons) and for the force-carrying bosons (e.g., gluons, photons, and W and Z bosons). The MSSM had these terms, plus two more, or equivalent to the quarks and leptons, except as bosons, and the other equivalent to the force carriers, except as fermions.

With the addition of these terms, a most unsettling thing occurred. Just like the existence of the first two types of particles—fermions and bosons—in the standard model meant that of necessity quarks, leptons, and the force-carrying bosons had to exist, adding the second two terms in the MSSM predicts that still more particles must exist. Quite logically the number of particles predicted by the MSSM is precisely double those we currently know about. That we've never seen these extra particles has led some skeptical physicists to note (with not a little sarcasm) that at least we've discovered half of the particles predicted by the MSSM theory. The MSSM also requires that we have not one, but four, different Higgs bosons. Thus, if the LHC experiments find more than one Higgs boson, this will be evidence that the idea of supersymmetry has some merit.

Simple naming rules apply to these newly predicted particles. The bosonic equivalents to our familiar matter particles would have the same name, except with an "s" before them. Thus quark would become squark; lepton would be-

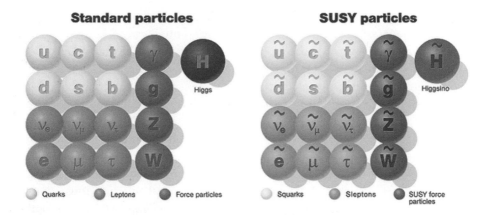

Figure 2.7. The supersymmetric, or SUSY, particles have an identical organization to that of normal particles. They are indicated by adding a tilde over the abbreviation of the particle. Courtesy of Deutsches Elektronen-Synchrotron.

come slepton, and so on. The fermionic equivalent of the familiar force-carrying particle would get an "-ino" added to the end, with occasionally a little phonetic surgery to make the word easier to say. So the W boson would become a wino (pronounced "ween-o"), photon would become photino, and so on. In all cases, we can denote a supersymmetric particle by putting a tilde (~) over the symbol, thus a photino is $\tilde{\gamma}$, a stop squark $\tilde{t}$. Figure 2.7 gives the entire list.

In a moment, we'll get to how we might find supersymmetry in the LHC's detectors. But in the meantime, let's discuss why you might want to add terms to your equations that would double the number of particles you predict, with zero physical evidence that they actually exist. As I mentioned at the beginning of this chapter, there are mysteries in the universe. And, no, I'm not talking about "What do women *really* want?" or "Why can't men keep the toilet seat down?" I'm talking about questions of physical phenomena. One of the interesting questions is that of force unification. Just like Newton showed that the phenomenon that keeps my cat firmly placed on my keyboard as I write is the same thing that governs the planets, current physicists hope to show that the four forces of which we are aware are really one and the same. It is not true that this idea is taken as an article of faith, nor do all physicists believe that it is inevitable. But it sure would be elegant if it were true.

In the 1960s, physicists were able to show that the weak and electromagnetic forces were one and the same (and should properly be called the "electroweak force"). Now the questions are, "Can we show that the strong force and the electroweak force are just different ways of looking at the same thing? And can we show that they are both the same as gravity?" The answer to these questions is currently no, but there are reasons to think that these questions are within the realm of the possible and that the answers may eventually become yes.

For instance, we can measure just how strong the three quantum forces (strong, electromagnetic, and weak) are. If we measure the strength at different energies (that is to say in collisions of different violence), we see that their strengths aren't constant and actually change with energy. If we project the trend of the three forces, we see that they all become the same at the rather high energy of 10^{14} to 10^{15} GeV, or a hundred thousand billion times the mass of a proton. We call this the grand unification theory of energy. Contrast this to the energy at which the symmetry between electromagnetism and the weak force is broken, which is about 1,000 GeV. The fact that the three forces "just happen" to have the same strength at some energy is suggestive (but not proof) that they are all one and the same thing. The fact that they all merge at one particular energy is very interesting.

On closer examination, we see that the three forces do not project to exactly the same spot in the standard model theory, as shown in Figure 2.8. However, and this is a suggestive beauty of the MSSM, it is pretty easy to use supersymmetry to make the three forces merge at *exactly* the same energy. This is not proof that supersymmetry is right, but it gives us a warm and fuzzy feeling nonetheless.

Even without the perfect unification of supersymmetry, there is a nagging question. Why does electroweak symmetry breaking occur at so much lower an energy than the grand unification energy? That's just weird and unnatural. It's kind of like the finances of a billionaire. Every month, she has earnings and expenditures. Large amounts of money slosh into and out of her bank account. If at the end of every month her account had under a dollar in it that would be weird. It's hard to imagine these large million-dollar deposits and withdrawals could balance so perfectly without some principle making it so.

If it turned out that this was her "charity account" and it was set up so that the deposits were carefully designed to cover planned donations that were automatically transferred to the charity, then the bank balance would make sense. But without the "charity principle," it would remain very mysterious that the account would be balanced so well.

Similarly in physics, it's rather odd to have the grand unification theory scale to be a hundred billion times greater than the electroweak symmetry breaking scale. By all rights, they should both be more similar (and nearer the high end). Thus the Higgs boson mass (recall this plays a critical role in electroweak symmetry breaking) should be much, much higher than suggested by the data we've collected to date. So perhaps there is a principle that explains this disparity in the unification energies, and supersymmetry seems to fit the bill

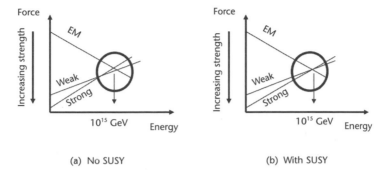

Figure 2.8. While all forces seem to approach the same strength at high energy even under the standard model (shown as No SUSY, *left*), it is possible using supersymmetric principles to make the three forces unify at a single energy (shown with supersymmetry, or SUSY, *right*). GeV = one billion electron volts; EM = electromagnetic force.

quite well. If supersymmetry turns out to be true, it's relatively easy to explain the low Higgs boson mass.

Even though we have yet to prove the concept of supersymmetry, we can make predictions about it. One prediction is that there should be twice the number of particles compared with what we currently know. So what do we know of these hypothetical particles?

The answer is very little. We know that we haven't found them, which means that they have to have a large mass, if they exist at all. This brings us to another point. Recall what we're talking about: supersymmetry, where the operative word is symmetry. If the symmetry of supersymmetry was, in fact, symmetrical, then the masses of the new particles would be the same as regular particles. We would see that the mass of the selectron was the same as the electron. But we don't. That means that just as the Higgs idea (or something equivalent) breaks the symmetry between electromagnetism and the weak force, something breaks the supersymmetric symmetry as well. That's another conundrum, about which we have some ideas. But until we start finding some new particles, this is a concern we can table for the time being.

Given that we haven't discovered these new supersymmetric particles, we can conclude that if they exist, they must have a mass no lower than about a hundred times heavier than a proton or more. (If their mass were lower, we would have found them already.) But what would we expect to see if they're real? The LHC is a proton collider and therefore mostly collides quarks and gluons, as they are natural constituents of protons. In a collision making supersymmetric particles, we would expect squarks and gluinos to be made most frequently as these particles also are predicted to feel the strong force.

Many different types of interactions are possible involving supersymmetric particles, but supersymmetry makes one useful and pervasive prediction. If a supersymmetric particle is made, then it always has to have a supersymmetric particle in its decay. That means that supersymmetric particles will decay until the daughter particle is the lightest supersymmetric particle, or LSP. Because supersymmetric particles must have a supersymmetric particle as a daughter in the decay and the LSP (by definition) is the lightest supersymmetric particle, there is no lighter possible supersymmetric daughter. Therefore, the LSP is stable.

Further, if the universe once made gazillions of supersymmetric particles, it should contain a similar number of LSPs from all the decays. Since we haven't detected them, we know the LSPs must be electrically neutral. Consequently, they can't interact via the electromagnetic force. The strong force is out, too, although the weak force is in according to the theory. So if supersymmetry is real, the universe should be full of LSPs, essentially a bath of heavy, neutrino-like particles. This idea has consequences for cosmology, which we return to in chapter 5.

Let's go back to the LHC and consider an event in which we make two squarks, like the one shown in Figure 2.9. The squarks must be heavy and decay into quarks and LSPs. The quarks make jets as usual, and the LSPs escape undetected as they don't feel the strong or electromagnetic force. So in this particular case, you'd expect to see two jets and you'd notice that energy is missing. Thus your detector would need to be able to measure jets and also to measure energy accurately enough to know some is missing. This is a crucial capability, as the one common prediction of essentially all supersymmetry-incorporating theories is the existence of an LSP that can escape the detector and thus that energy will be missing.

There are many types of collisions possible in the supersymmetry theory, more than we can discuss here. Even worse, we recall that supersymmetry is a *principle* and not a theory in its own right, so this means that there are many theories that include supersymmetry. Most of these theories are much more complex than the relatively simple MSSM. So killing the idea of supersymmetry will be hard. The best we can do is to disprove individual models that incorporate supersymmetry. However, we do know one thing. If (and I stress the *if*) supersymmetry is the explanation for why the Higgs boson mass is so low, then we must be able to find supersymmetric particles at the LHC. If we don't, then we may not have entirely killed the principle of supersymmetry, but we will have ruled it out as an explanation for the light Higgs boson mass.

Just to make things a bit more interesting, some models that incorporate supersymmetry make predictions about the Higgs boson itself. Some of the models actually predict that there is more than one Higgs boson and even that the Higgs bosons carry an electric charge. This is clearly a very different predic-

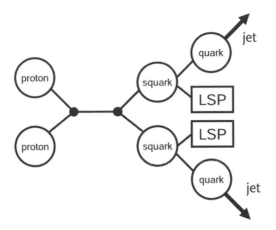

Figure 2.9. Representative decay of supersymmetric particles. Supersymmetric particles decay until they end with the lightest supersymmetric particle (LSP) daughter. The quarks turn into jets, which are sprays of particles in the detector. The solid circles denote a spot at which an interaction occurred.

tion than standard Higgs theory and is therefore useful in helping us to sort out what is right.

The LHC and its detectors were designed with all these questions in mind. It should be able to resolve some of this theoretical controversy. Those first few years after start-up will be unbelievably exciting, something that happens only three or four times per century.

Smaller than Small

The Victorian-era mathematician Augustus de Morgan wrote:

> Great fleas have little fleas upon their backs to bite 'em,
> And little fleas have lesser fleas, and so ad infinitum.
> And the great fleas themselves, in turn, have greater fleas to go on,
> While these again have greater still, and greater still, and so on.

This oft-quoted passage is a parody of Jonathan Swift's 1733's *On Poetry: A Rhapsody.* Although this was written about poetry, scientists have taken those lines as a metaphor for the natural world. As one learns about the microworld, one is quickly faced with the observation that all matter is made of molecules. Molecules are in turn made of atoms, which are themselves made of electrons and atomic nuclei. The nuclei are made of protons and neutrons, and these are composed of quarks. This progression to ever-tinier structures is illustrated in Figure 1.4.

However, as far as we know, quarks and electrons are it. That's the end of the line as far as structure goes. Unlike the atom or the proton, both of which have a rich structure with complex interactions between their components, quarks and electrons are currently believed to have no internal structure at all. Both theoretically and physically, they are considered to be mathematical points.

Of course anyone with an ounce of imagination can't help saying, "Now just hold on a minute. Why couldn't the quarks and leptons themselves have structure?" Well there's only one possible answer and it is, "They could." The quarks and electrons (and, by extension, all leptons) could be made of even smaller objects. Or they (rather improbably) may indeed be fundamental (i.e., structureless). In the following pages we consider the evidence for structure as well as how we might winnow out the answer to that question.

Before we proceed farther, let's consider the sizes involved. For convenience, they are tabulated in Table 2.1. Everything in the microworld is small. A single molecule is so small that you could lay a million of them side by side in a single millimeter. They are so small that you can't use ordinary light to see them. And yet, such smaller objects are enormously large; they are a billion times larger than the sizes explored at the research frontier.

Molecules are composed of atoms, which are about a tenth their size. This factor of ten is not very precise, as there are many kinds of molecules, from hydrogen, consisting of just two hydrogen atoms (H_2), through simple sugar, with 24 atoms ($C_6H_{12}O_6$), to large organic molecules, consisting of hundreds of atoms. However, we can roughly say that a millimeter is ten million times larger than an atom.

The mental picture of an atom as a little solar system, with a nuclear sun and planetary electrons, is flawed and yet not without merit. It highlights the fact that an atom consists of mostly empty space, with the electrons swirling frenziedly far from a small, dense nucleus. Figure 1.2 (and the relevant discussion in text near the figure) gives an idea of the relative sizes involved. Most important for our purposes, this figure shows just what a tiny fraction of the atom the nucleus takes up. The radius of the nucleus is about ten thousand times smaller than the atom, and the nucleus takes up but a trillionth of the volume.

The nucleus of the atom consists of protons and neutrons, packed tightly together. My mental picture of the nucleus is a mass of frog eggs, or marbles after being handled by a toddler with very sticky fingers. Each proton or neutron is about 10^{-15} meters wide and you would need a trillion laid end-to-end to span a single millimeter. That's small.

Protons and neutrons contain within them quarks and gluons. The simplest way to think of a proton is that there are two up quarks and one down quark

Table 2.1 The enormous range in size of the supersmall

Object	Size (meters)	Size (relative to a proton)
Molecule	10^{-9}	1,000,000
Atoms	10^{-10}	100,000
Nuclei	10^{-14}	10
Protons	10^{-15}	1
Quarks and electrons	$< 10^{-18}$	< 0.001

stuck in a force field of gluons. Think of three Ping-Pong balls in one of those air-blown lottery machines and you get the basic idea.

The mental picture of quarks as Ping-Pong balls has one major drawback. Ping-Pong balls are large compared with a lottery machine. Quarks are small. So maybe a better mental picture of the proton is three little flecks of Styrofoam in the same machine.

So what do we know of the size of quarks? Earlier I said that they have no size, and that's certainly how the current theory treats them. However, as an experimenter, I'm more concerned with measurements. You the reader must be curious as to what measurements have revealed the size of a quark to be. And now the answer . . . a drum roll, please . . . they haven't. This doesn't mean we know nothing. We've studied this question rather thoroughly and we know very precisely how good our equipment is. If quarks (and electrons) were larger than about a thousand times smaller than a proton, we'd have seen that they have a size. In all of our experiments, we've never seen even the slightest believable hint of a size. We therefore conclude that while we can't say what the size of a quark or electron actually is, we can safely say that if quarks have size at all, they are smaller than a thousand times smaller than a proton.

If this idea is hard to understand, let's consider how small an object you can see with your eyes. You can easily see a grain of sand. With very considerable effort, you might be able to see the smallest bit of flour in your cupboard. But that's about it. With your bare eye, you can't see anything smaller. Thus when you decide to look at a germ with your eye, you could conclude that they have no size, but the strictly correct conclusion you should draw is that germs are smaller than a tiny fleck of flour.

With better equipment, say a powerful microscope, one can see that germs actually do have a measurable size. So once you've hit the limitation of your equipment, you simply need to get a more powerful microscope. The microscope that is the LHC and its two primary detectors will observe the size of quarks if they are no less than one ten thousandth of the size of a proton; in other words, they will set a limit that is about ten times smaller than is currently thought.

So, observations, intuition, and de Morgan's ditty may be enough to reveal a casual suspicion that there may be other levels of matter that occur at ever-smaller sizes—a whole new layer or set of layers in the cosmic onion. But there are more scientific reasons as well. For instance, consider the periodic table shown in Figure 1.1. While Mendeleev intended it to be an organizational scheme, with the formulation of the theory of the nuclear atom and quantum mechanics in the first few decades of the twentieth century, it became clear that the periodic table was actually the first indication of atomic structure, half a century before this structure was truly understood.

To make this point more clearly, let's focus on the columns at both ends of the periodic table. The leftmost column includes chemically active elements. Hydrogen, lithium, sodium, and all the rest are chemically similar and have the same valences (for those of you who recall your introductory chemistry classes). Yet each of these elements in turn is heavier than the one above them in the column. With our understanding of chemical structure, we came to understand the increasing mass as being caused by ever more protons and neutrons in the nucleus, while the chemical similarity turned out to be explained by a repeating pattern in the arrangement of atomic electrons, with each of these elements having a single electron available to interact with other atoms. These atoms have differing numbers of electrons, but all but one of them are safely packed away, unable to interact with other atoms.

The story in the right-hand column is essentially identical. Helium, neon, and argon are all chemically similar elements with increasing mass. They are all inert because of the arrangements of atomic electrons. These elements all have their electrons tidily packed away around the atom, with no stray electrons available to interact with other atoms.

The story told by the periodic table clearly hinted at atomic structure. Similiarly, the story of nuclear radiation suggests a nuclear structure. For instance, cesium ($^{137}_{55}$Cs, with 55 protons and 82 neutrons) emits an electron and becomes barium ($^{137}_{56}$Ba, with 56 protons and 81 neutrons). This decay emits a neutrino, too, although that fact was not definitively established until the 1950s. This decay could be understood as having a neutron in the cesium spit out an electron and thereby became a proton. But even before protons, neutrons, and neutrinos were understood, the idea that the nucleus of one element could change into the nucleus of another element was seen as a hint of nuclear structure.

So let's take these historical examples and apply the reasoning to the modern world. We realize that historical lessons do not always apply. But sometimes they do.

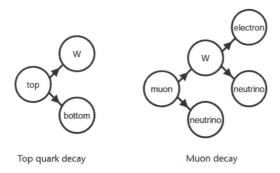

Top quark decay   Muon decay

Figure 2.10. Decay of particles that transform from one quark or lepton type to another.

Our current periodic table is shown in Figure 1.6. Its organization is different from the chemical periodic table. In the figure, there are six types of quarks. The up, charm, and top quarks all have $+\frac{2}{3}$ charge (in a system where the charge of a proton is +1), and the mass of the charm quark exceeds that of the up quark, which in turn is surpassed by the top quark. Similarly, the down, strange, and bottom quarks all have electric charge $-\frac{1}{3}$, with the mass increasing as one goes toward the right.

In the case of the leptons, the electron, muon, and tau all have an electric charge of -1, with the usual mass pattern. The three neutrinos are all electrically neutral and their mass is not known, although the fact that they have nonzero (and different) mass is not in dispute.

In the modern periodic table, the chemically similar units are the rows, in contrast to the columns of Mendeleev's table. We see that there are three generations, or carbon copies, of the same quark and lepton pattern. This is highly reminiscent of the hints that the chemical periodic table was giving us in the second half of the nineteenth century.

There is another historical similarity to consider. Just like the various atomic nuclei could decay into other nuclei, so too can the quarks and leptons. A top quark can decay into a bottom quark and a W boson. Likewise, the muon can decay into an electron and two neutrinos. These processes are sketched in Figure 2.10. Other types of quark and lepton decay are also possible. In fact, all particles in generations II and III eventually decay into the particles of generation I. One crucial clue is that the only force that can change one quark or lepton into another (we say "change the quark or lepton's 'flavor'") is the weak force. Further, only the electrically charged W boson can do the job.

So while there is no hard evidence that the presence of quark and lepton generations indicates that quarks and leptons are themselves composed of smaller (and thus far undiscovered) particles, the historical analogy is powerfully suggestive and certainly warrants closer attention. That, by emitting a W boson, one can change the quark or lepton flavor is an extremely valuable clue that is screaming something important at physicists.

I just wish that I had the wits to understand what it was saying.

Even without the crucial insight that cracks the conundrum wide open, however, we can speculate intelligently on the subject and (much more important) sift through our mounds of data, looking for additional clues. As with all searches for new physical phenomena, you have to make an educated guess about what to look for and then look for it. So what are the likely experimental signatures of quark or lepton structure? Since the LHC will be colliding protons (which are essentially bags of quarks), we focus on the search for quark structure.

Before we get into specifics of quark structure, let's consider how a hypothetical being the size of a galaxy would prove that the Earth had a size and isn't a single point of matter. Recall that Newton's law of gravity treats all objects, even planets, stars, and galaxies, as pointlike particles. As long as you are outside a star, you can replace an entire star with the same amount of mass concentrated at a microscopic point and not be able to tell the difference at least as far as gravity is concerned. Once you got inside the star, then the rules would change and the two cases (the star's having a size and there being a pointlike mass) would not be equivalent.

Let's go back to our method of deciding whether the Earth has a size or not and consider Figure 2.11. The way a galactic-sized being might figure this out would be to take comets and fling them toward Earth. As long as the comets don't pass closer than 6,400 km (4,000 miles) from the center of the Earth (note that the radius of the Earth is the same measurement), he can't distinguish between the real Earth and the pointlike Earth. All comets' paths would be bent identical amounts by gravity.

When the comets are made to pass within 6,400 kilometers (4,000 miles) of the Earth's center, well, then the two models would give different results. For the pointlike Earth equivalent, Newton's laws would still apply, and you could use them to calculate how the comet would be affected by gravity. The real Earth would act differently. The comet would plow into the Earth's surface, and different physical principles would apply. The electromagnetic force that governs the behavior of the atoms in the comet and the Earth would determine how big the "splat" will be. So at a particular size the relevant laws of physics change, with gravity no longer being the only relevant force and electricity (i.e., atomic forces) taking over.

There is an analogy that illustrates how the beam energy—in other words, the beams used to collide protons in the LHC or any other particle accelerator—

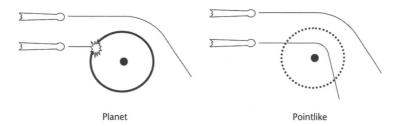

Planet   Pointlike

Figure 2.11. As long as a comet's path takes it outside the planet's surface, a planet (*left*) and a point-like mass (*right*) are indistinguishable from the comet's point of view. However, once the comet's path brings it to a radius smaller than the Earth's surface, you can tell the difference.

matters a lot in the ability to see the supersmall. To illustrate this idea, we need to recall a couple of things. The first is the famous quantum mechanical postulate that objects can act both like particles and waves. It is the wave nature of our particles that is relevant here. For purposes of our discussion, we need to consider two principles: wavelength and diffraction.

Wavelength is the distance between peaks in a wave. The wavelength of a particle is related to the energy of the particle. The higher the energy, the shorter the wavelength, as illustrated in Figure 2.12. So why is the wavelength relevant? This is where diffraction comes in. Suppose you're looking at a lake in which there are waves moving past a stick stuck in the water to measure water level. If the waves are very long, they move past the stick without any notice, unaffected by its presence. However, if the waves are short, as shown in Figure 2.13, there is a "shadow" of the stick after the waves pass by.

The important points here are the following: (1) to see something small requires wavelengths even smaller than the object being observed, and (2) particles with high energy have short wavelengths. In fact, the wavelength of a particle with the full energy of the LHC is 2×10^{-19} meters, or about ten thousand times smaller than a proton. This means that if the quarks and leptons have a size slightly greater than this number, then the high energy beam of the LHC will be able to distinguish between a pointlike particle and one with a size. In the next chapters we will look much more closely at how the LHC, and particle accelerators in general, works.

With all these preliminaries out of the way, you might ask, "OK, but what will physicists at the LHC be looking for that could signal quark structure?" Several techniques will be used. As with all frontier research, we don't know what the answers will be and therefore we will look in a lot of places. One of them may (and, as usual, I stress the *may*) be the right place to look.

Historically, one of the best places to look has been the most violent collisions. You smash two objects together and see how many collisions there are

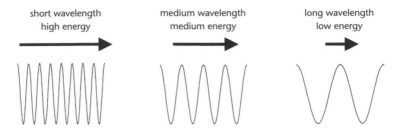

short wavelength
high energy

medium wavelength
medium energy

long wavelength
low energy

Figure 2.12. The effective wavelength is related to the energy of a particle, with lower energy particles having longer wavelengths. Seeing small objects requires short wavelengths and therefore large energies.

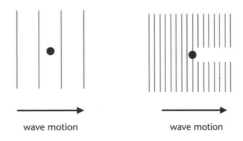

wave motion

wave motion

Figure 2.13. An illustration of how waves interact with a stick in the water. If the wavelength is larger than the object, the waves pass by unscathed. If the wavelength is smaller than the stick, the waves will be disturbed as they pass by it.

at each level of violence. Specifically, you look at the amount of "sideward violence." Technically we call this "transverse momentum," which means perpendicular to the beam. There are technical reasons for this choice, but mostly it is because you have to hit something hard for it to go sideward from its original direction.

Let's examine at what an experimental signature of quark structure might look like. We recall from our earlier discussion of the Higgs boson that if you smash a quark out of a proton, it forms a jet. We can simply add up all the energy of the particles in the jet and that does a pretty good job of looking a lot like the original quark. So we'll just talk about quarks here, although experimentally we measure jets.

In Figure 2.14, we see a plot of how often a collision of a particular level of violence occurs. First, we see that low-violence collisions are more likely. Looking at the region of energy labeled "proton-proton regime," that's where the protons are collided with such little energy that the protons act as little billiard balls, and there is no hint of the existence of quarks. The dashed continuation of the line shows what the theory predicted would happen if the protons had no

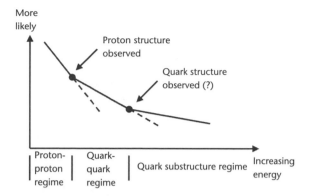

Figure 2.14. The correspondence between the violence (or energy) of the collision and how likely it is. At low energy, protons collide with such little violence that they don't break up. However, when these collisions become more energetic, the protons begin to fall apart and the quarks contained within them start to be seen. The onset of seeing quarks comes with a change in the slope in the graph. The dashed line indicates the behavior we expected to see if protons had no quarks within them. As protons are collided more violently, we hope to see objects inside quarks using the same techniques. Another change in slope will hail the onset of seeing quark substructure.

structure and always acted pointlike. However, at a particular level of violence, the decreasing trend changes. This is because at that level of energy, the collision became violent enough to see individual quarks, rather than the proton as a whole.

Harking back to our earlier analogies, you can think of the protons as finally running into one another or the energy finally being high enough to make short wavelengths. (Indeed, the technical answer includes both ideas.) In any event, the signature that demonstrated the existence of quarks was that at a particular level of violence, there started to be more of that particular energy collision than you would expect from the lower-energy trend. In Figure 2.14, we predicted we would see the number of collisions indicated by the dashed line, but what we actually *saw* was the solid one. Quarks were discovered using this and other techniques in the 1970s and early 1980s.

We expect the case to be similar in the event that quark structure is observed at the LHC. Because the energy in the LHC's beams of protons is unprecedented, perhaps we will finally make collisions with sufficient violence to start seeing more than the long-observed trend in the scattering of quarks.

Scientists have many ideas as to what might be found inside quarks (including the idea that quarks are indeed pointlike). While the "up-like" quarks have a charge of $+\frac{2}{3}$, "down-like" $-\frac{1}{3}$, and leptons -1, the objects within quarks could have charges that are a multiple of $\frac{1}{3}$, $\frac{1}{6}$, or other possibilities. Unlike the case of electroweak symmetry breaking and the Higgs boson, no favorite has emerged from the various contenders. Indeed, with precisely zero direct evidence for the existence of quark and lepton structure, most physicists have taken a "wait and see" attitude, preferring to see what hints the universe will give us. Even so, names have been proposed for these objects smaller than quarks, with the most popular being "preon" (for pre-quark). However, each theoretical physicist who has devised a theory has invented his or her own name, with subquarks, maons, alphons, quinks, rishons, tweedles, helons, haplons, and Y-particles all having been suggested. I kind of like the names quinks or tweedles myself.

One additional question on the topic of quark structure is the following: Just like the atomic model of the atom has led us to find more and more elements, would it not be true that we would expect additional generations of quarks and leptons? Why are there only three, and how do we know there aren't four or more?

The short answer is, of course, there could be more generations. Experiments have tried to find the so-called b-prime quark, which is a yet unnamed fourth generation "bottom-like" quark. As of the summer of 2008, no evidence for its existence has been observed.

Probably the strongest evidence for there being only three generations comes from the LEP experiments that once inhabited the underground tunnel that now houses the LHC. The LEP accelerator collided electrons and positrons in the heart of four superb detectors. Much of the time, the detectors tuned their beam energy to make millions of Z bosons. With such a large sample of Z bosons, experimenters could study them in great detail, and the precision of these measurements is extraordinary. The LEP experimenters were able to measure the fraction of time a Z boson decayed into quarks, electrons, muons, and so on.

For purposes of our discussion, the interesting decay was when the Z boson decayed into a pair of neutrinos. In Figure 2.5, we noted that the Z boson decayed into neutrinos about 20% of the time. Since neutrinos don't interact with matter, these kinds of decays are never observed. However, they do make their presence felt. One can calculate how often you expect the LEP beams to make a Z boson depending on how many neutrino generations there are. The LEP experiments concluded that the data were consistent with there being between 2.95 and 3.05 generations. Because the only possible answers for the number of generations are integers (1, 2, 3, 4, . . .), that's just a fancy way to say that LEP concluded that there were precisely three neutrino generations with considerable certainty.

So that's pretty strong evidence that, for whatever reason, the universe allows only three generations. However, there is one little bit of wiggle room. Technically, the LEP experiments showed that there were three generations of nearly massless neutrinos. If there is a fourth generation and the neutrino of this generation is massive, then the LEP data can't rule that out. While there is no reason to expect a generation IV neutrino to be massive, that the top quark is so much more massive than the other quarks tells us that the idea of a hypothetical heavy neutrino isn't ludicrous. After all, there are ample examples of particles in higher generations being more massive than their lower-generation counterparts, so the idea remains viable. Like anything at the research frontier, only through experiments will the question of quark and lepton structure be resolved.

What's the Matter with Antimatter?

"Space. The final frontier" is the opening of a wonderful television show from the 1960s called *Star Trek*. A youthful Don would look forward to watching it in syndication, peering at a fuzzy picture on a UHF station. (You youngsters ask your parents what UHF was. If you know, don't admit it, because that means you're getting old.) In this show, a mighty starship called the U.S.S. *Enterprise*, captained by the legendary James T. Kirk, would scoot around the galaxy encountering situations that frequently had moral relevance to the social problems of the day.

For our purposes, the show itself isn't so important as the ship's engines. They were powered by antimatter. It could be true that, like many of the high-tech doodads in the show, antimatter was merely a fictional device, on par with dilithium crystals, a convenient futuristic plot device to make plausible their speedy journeys.

However, unlike many things that appear in science fiction, antimatter is entirely real. (Further, antimatter could be used, as it is the highest energy power source ever discovered, making not so silly its presence in a starship's engines.) Antimatter is the opposite of matter and, when combined with matter, will completely annihilate into pure energy.

Antimatter is perhaps most simply understood at a particle level. For every particle discovered, there is a corresponding antiparticle. There are antimatter electrons (with the special name of positron) and antiquarks, which have no special name. Some particles, like the photon, are their own antiparticle. From antiquarks, you can make antiprotons and antineutrons. Toss in antielectrons and you can make antiatoms. With antiatoms, you could in principle make anti-anything: anti-you, anti-me, antipasto (or should that be anti-antipasto?), and on and on.

Antimatter can be created in physics laboratories by converting a prodigious amount of energy into matter and antimatter. In fact, it would take the entire energy of a Hiroshima-type nuclear explosion, converted with 100% efficiency, to make enough antimatter to make an anti-paper clip.

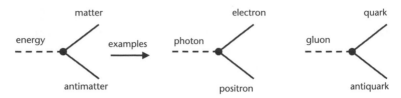

Figure 2.15. Examples of how energy can be turned into antimatter but only with a corresponding matter particle.

The operative words here are "into matter *and* antimatter." The way anti-matter is made is to convert energy into *identical pairs* of matter and antimatter. Figure 2.15 illustrates the general idea. Yet this leads us to one nagging problem. If we combine two observations, namely (1) matter and antimatter are made in equal pairs, and (2) we only see matter in the universe, we're led to the obvi-ous question "Where the heck is all the antimatter that should be here?" This remains one of the unsolved mysteries of science.

We do know some things about how the universe treats matter and antimat-ter. To the best of our knowledge, the strong and electromagnetic forces treat matter and antimatter identically. However, the weak force doesn't treat them the same, as was shown in a series of experiments beginning in 1956.

Before I describe the outcome of these experiments, let's backtrack a little and introduce a few more particles. First, this book is about the "Large Hadron Collider," so what is a hadron? It is a subatomic particle, such as a proton or a neutron, that is composed of quarks. There are two types of hadrons: baryons and mesons. Baryons are made up of three quarks, and mesons are made up of a quark and an antiquark. Mesons are further divided into different types, includ-ing pions (made up of a quark and an antiquark of the up or down variety and is shown with the Greek letter π), neutral K mesons (which contain either a strange quark and a down antiquark or a down quark and a strange antiquark), and neu-tral B mesons (containing a bottom quark and a down antiquark or vice versa).

In 1964, a paradigm-shifting experiment revealed a slight asymmetry in the decay of neutral K mesons (or kaons) and gave the first indication that matter and antimatter might act slightly differently. Later, in 1999, additional measure-ments involving kaons revealed more about the matter-antimatter asymmetry. Knowing that some physical processes favored matter over antimatter was a huge step in understanding why we live in a matter-dominated universe, but the slight preference for matter observed in neutral kaon decays wasn't enough. There had to be more. Even several decades ago, physicists calculated that they expected a greater asymmetry in mesons involving bottom quarks. The story of

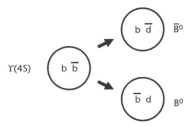

Figure 2.16. The ϒ(4S) particle, containing a bottom quark (b) and antiquark (b̄) can decay into pairs of neutral bosons called mesons containing bottom quarks. These bosons are the B̄⁰, containing both bottom and antimatter down quarks (d̄), and the B⁰, in which the quark and antiquark are reversed. This is an especially interesting process for studying the matter-antimatter asymmetry.

the study of neutral kaons should not be treated this cursorily, because the story involves brilliant scientific detective work. But to limit the scope of this book and to focus on the LHC, the study of neutral kaons is only sketched here. The full story is given in the suggested reading.

In 1999, two detectors, Belle in Japan and BaBar in California, were turned on with the sole purpose of making bottom-quark-containing mesons in prodigious quantities, with neutral B mesons being of particular interest. Because the study of hadrons containing bottom quarks is an important goal for the LHC, we will sketch the Belle and BaBar experiments here. While scientists working with Belle and BaBar had intermediate successes, in 2004 they announced an enormous asymmetry that preferred matter over antimatter in the decay of neutral B mesons. This preference for matter is one hundred thousand times greater than that seen in the decay of neutral kaons discussed above.

Both Belle and BaBar were detectors designed to study collisions of electrons and positrons. And not just any old collisions would do; the beam energies were carefully selected to produce the so-called ϒ(4S) meson (that's "upsilon four S"), which consists of a bottom and antibottom quark. The ϒ(4S) can decay into two neutral B mesons. These daughter mesons are called B̄⁰, consisting of a down antiquark and a bottom quark, and B⁰, consisting of a bottom antiquark and a down quark. You'll note that this is similar to the neutral kaon case, with the bottom quark taking the place of the strange quark. Figure 2.16 shows how the ϒ(4S) can decay, highlighting that matter bottom quarks (denoted b) and antimatter bottom quarks (denoted b̄) occur in equal quantities.

To study the decays of neutral mesons containing bottom quarks, a particular decay mode was used. This decay was into a charged K meson and a charged π meson. Because the B meson was electrically neutral, the K and π mesons had to have opposite electric charges to balance out. These mesons consisted of the

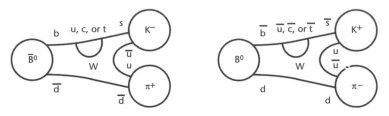

Figure 2.17. An illustration of how neutral B mesons can tranform into charged π and K mesons. Most readers can ignore the details of the transformation and should focus only on the fact that the sign of the charged K meson identifies the parent neutral B meson.

following quarks: K⁺ (up and anti-strange), K⁻ (strange and anti-up), π⁺ (up and anti-down), and π⁻ (down and anti-up). The need for this level of detail will become apparent in a moment.

Figure 2.17 shows how B^0 and $\bar{B}^0$ mesons decay. Most readers can ignore everything between the circles on the ends. The most important thing to note is that if you see a K⁺ π⁻ decay, you know it came from a B^0. A K⁻ π⁺ decay comes from $\bar{B}_0$.

So here's the big point. Because B^0 and $\bar{B}^0$ mesons are made in equal quantities, you'd expect to see the K⁺ π⁻ and K⁻ π⁺ decay modes occurring with equal frequency. But both Belle and BaBar didn't. Both saw that there was about 80% of the K⁺ compared with K⁻. That means the number of antimatter bottom quarks seen was about 80% that of bottom quarks: less antimatter than matter.

This measurement firmly established that the study of bottom quarks could easily be crucial for shedding light on the preponderance of matter in the universe. In both Belle and BaBar, these measurements took about 200 million Υ(4S) and ended up with about a thousand charged K and π decays. A thousand samples of the desired decay isn't all that many, highlighting the need for more data. In addition, the decay mode described here is only one of many being investigated. Thus many more examples of bottom-quark-containing hadrons can only help the situation.

The LHC is, in many ways, a superior source of bottom quarks compared with Belle and BaBar's electron-positron collider. Belle and BaBar's strength was the fact that they had an exquisitely pure sample of specific bottom-quark-containing hadrons. Unfortunately, this fact is also their weakness. They can make what they make and that's it. In contrast, different types of quarks will abound in the LHC. Many different hadrons containing bottom quarks are possible, allowing for a much richer set of studies. These studies are a crucial point of the LHC's design and research goal.

Heavy Ions

Most of the LHC's experimental program will focus on colliding protons together at the highest energy; however, this is not everything the LHC will do. About one month each year, the LHC will accelerate the nuclei of atoms of the element lead and collide them together. The physics questions being explored in lead-lead collisions are vastly different than those in proton-proton ones. In proton collisions, the idea is to focus as much energy in as tiny a spot as possible, like the pressure at the point of a pin. In the case of lead collisions, the idea is to spread a lot of energy over a large volume. Energy across a large volume can reveal phenomena that a focused energy collision would miss.

Like all experimental programs, one can ask countless different questions and make countless different measurements while colliding lead nuclei. However, there is one phenomenon that stands out in the study of lead and indeed all heavy nuclei collisions. This is observing and characterizing an entirely different type of matter.

People are familiar with the three most common states of matter: solid, liquid, and gas. When you think about it, the insight that the same materials can have such vastly different properties and yet still be the same thing is pretty amazing. Air and a frozen clod of dirt (i.e., a gas and a cold solid) are entirely different things and yet steam and ice (also a gas and a cold solid) actually *are* the same thing.

We call these various states of matter "phases." What most people don't know is that there are other phases, both observed and merely hypothesized. Typically one can change matter from one form to another by heating or cooling it (or equivalently adding or subtracting energy). Let's think about what happens to water when you add energy to it. Start with a familiar ice cube. If you heat it, first the ice cube warms up. When it reaches 0°C (32°F), the ice melts. Water changes from its solid to its liquid form, that is to say it changes phase. Heating the water changes it to steam, water's gaseous form. In its gaseous form, individual water molecules can fly around, interacting very little with one another. In contrast, water in its liquid form exhibits very different behavior. Molecules of liquid water "know about" each other. That's why liquid water can experience such behaviors as viscosity and surface tension. The fact that the same matter can act so differently under different energy and temperature conditions is one of the reasons we study matter. We want to see all the rich behaviors that it can exhibit.

Since gaseous water consists of individual molecules, we need to know about them. Water molecules consist of three atoms, two of hydrogen and one of oxygen. (Hence, water's molecular formula: H_2O.) In its gaseous form just above the temperature at which water boils (100°C, or 212°F), each molecule acts individually.

However, as the temperature of the steam is increased, the water molecules bounce around with more and more energy. Eventually they start bouncing into one another hard enough that the molecules are broken apart, with hydrogen and oxygen atoms individually wandering around willy-nilly.

The electrons are strongly bound to their respective nuclei, but as the temperature is raised, the nuclei are no longer able to hold onto their electrons, which are then stripped away. Oxygen and hydrogen nuclei are intermixed with free electrons. The whole mix is electrically neutral. This is actually considered a new stage of matter called a plasma. You can see an example of an electrically produced plasma in a fluorescent light bulb.

Further heating this mixture eventually will cause nuclei to break apart, leaving electrons, protons, and neutrons flying around. Temperatures as hot and energies as high as this have been achievable for decades.

We recall that protons and neutrons are made of quarks. Each proton could be thought of as roughly a bag with three quarks locked firmly within it. Technically, as was discussed in chapter 1, we say that the quarks are "confined" in the proton or neutron. The question to be asked is whether at high enough temperatures can the quarks be freed from their protective nucleonic cocoons?

Figure 2.18 illustrates the basic idea. Above a certain temperature, the quarks are unconfined and allowed to mix freely. In some ways, the situation is analogous to a tumbler full of ice cubes (the protons and neutrons in the analogy), which melt when heat (i.e., energy) is added to form liquid water (the intermixed quarks). This state of matter has historically been called "quark-gluon plasma" in analogy with the electrical plasma of fluorescent lights.

The ice cube analogy is more apt than you might think. For a long time, it was thought that a quark-gluon plasma would behave like a superhot gas, with the quarks bouncing around, ignoring one another. However in 2005, experiments at the Relativistic Heavy Ion Collider, or RHIC, on Long Island, showed that when nuclei of gold collide, the resultant is a new form of matter that acts more like a liquid. In fact, when nuclei are heated to the point where the protons and neutrons melt in place, the freed quarks and gluons act like a liquid with zero viscosity.

Viscosity is a property of liquids that basically relates to how thick they are and how much they slosh. Take out a spoon, stir your coffee, and remove the spoon. The coffee will continue to swirl around in the cup. Repeat these actions in a bowl full of warm honey and the swirling will die down quickly. We say that honey is more viscous than water. Rather surprisingly, when quarks were freed

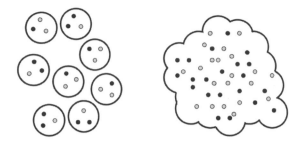

Figure 2.18. In ordinary matter (*left*), quarks are held inside protons and neutrons, three at a time. In a quark-gluon plasma (*right*), the quarks are no longer held inside the nucleons and are allowed to intermix freely.

from the proton and neutrons, the resultant state of matter acted like a fluid that swirled forever. It had zero viscosity.

So just how does one free the quarks? By heating up a nucleus, of course. Specifically, one aims beams of atomic nuclei at one another. These nuclei generally miss one another or occasionally experience a grazing impact. However, once in a great while, the two nuclei hit head-on. Just like hitting two bullets together at high speed can cause them to melt, doing so causes the nuclei to melt. (If you have trouble believing that an impact can make something warm, try banging a hammer many times on a solid rock and then feel the head of the hammer. It will be hot.)

Figure 2.19 shows an example of such a collision. In Figure 2.19a, the two nuclei are coming together at high energy. While nuclei are basically spherical, in a particle accelerator, they look more like two pancakes hitting face on. This is because of Einstein's theory of special relativity, which says that fast-moving objects will contract in their direction of motion. Thus the sideward dimensions are not shrunken and remain circular. The result is this pancake.

In Figure 2.19b, the "pancakes" pass through one another, with some of the energy being deposited in the nuclei, heating them up. If the conditions (i.e., energy levels) are right, the nuclear matter will be heated enough to free the quarks and a quark-gluon plasma will be formed as seen in Figure 2.19c. Finally, the ensuing fireball will continue to expand and cool off, with the quarks recoalescing into protons, neutrons, and other hadrons, as shown in panel 2.19d. The collision is over.

A fair question one might ask is this: "How would you know a quark-gluon plasma if you saw one?" Like every question asked at the LHC (or any modern experiment for that matter), there will be many different ways to answer. However, we can discuss one way that will illustrate the important points. This is a technique called *jet quenching*.

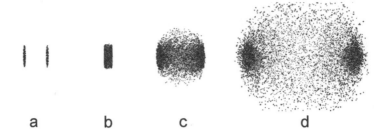

a   b   c   d

Figure 2.19. The stages of formation of a quark-gluon plasma: *a*, two nuclei approach one another, flattened into a pancake shape by relativistic effects; *b*, the actual collision; *c*, the formation of the quark-gluon plasma as the shock wave heats the nuclear matter; and *d*, eventual expansion and cooling of the fireball. Courtesy Jeffery T. Mitchell, Brookhaven National Laboratory; simulation by the UrQMD Collaboration.

Jet quenching is a pretty cool and intuitive idea. Earlier in our discussion of the experimental signatures of the Higgs boson, we mentioned that if you knock a quark out of a proton or neutron, a jet will form. Jets, we recall, are "shotgun blasts" of particles that are the characteristic signature of a quark escaping a proton.

In a collision between any heavy nuclei (e.g., lead or gold), two quarks will sometimes hit one another just like when beams consisting only of protons collide. In the case of a collision that is violent enough to dislodge quarks from a nucleon, but not violent enough to form a quark-gluon plasma, the scattered quark can escape the fireball mostly unscathed. The quark passes by the protons, neutrons, and other hadrons in the fireball. Because these hadrons have no net color (the strong force discussed in chapter 1), effectively the quark simply doesn't "see" them. This behavior is illustrated in Figure 2.20.

However, if the fireball is hot enough to melt the protons and neutrons into a quark-gluon plasma, then the violently scattered quarks will pass by free quarks, each with its own color. These high energy quarks will then bounce into the quark-gluon plasma quarks and sometimes never make it out of the fireball. So you'd expect to see that, as the collision becomes hotter and hotter, you'd see fewer and fewer high-energy scattered quarks (and therefore fewer jets). Thus we say that jets will be "quenched." This is one of the many signatures for which physicists will look.

So why is this kind of study interesting? It's because scientists believe that collisions like these re-create the conditions of the universe just instants after its creation. This is not a book on cosmology, but briefly we believe that the universe was once much smaller and hotter. This is what is known as the big bang

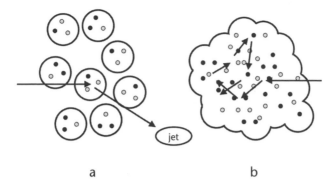

Figure 2.20. Jet quenching. In ordinary nuclear matter (*left*), a quark will scatter and leave the volume, making a jet. In a quark-gluon plasma (*right*), the quarks will bounce around, hitting the free quarks. Thus the quark does not leave the volume as often, with fewer jets being produced as a result.

theory. Much of the details of the very early universe are unknown, and only through intelligent speculation can we guess what might have occurred.

We believe that the universe once had a temperature that was high enough that the weak force and the electromagnetic force acted the same. Above this temperature, we say that the electric and weak forces were symmetric, and the study of the breaking of this symmetry—the concept of electroweak symmetry breaking discussed earlier in the chapter—is one of the important goals of the LHC. No matter the mechanism that breaks this symmetry, be it the Higgs idea, or something else, it is thought that the universe cooled enough that the electric and weak forces became two distinct phenomena by about a trillionth of a second after the big bang.

For the period of about one trillionth of a second until a millionth (10^{-12}–10^{-6}) of a second, the quark-gluon plasma is thought to have reigned supreme. The entire universe was so hot that quarks could (at least in principle) swim from one side of the universe to the other, unfettered by such considerations as protons and neutrons. It is this period in the history of the universe, called the "quark epoch," that the study of heavy ions is intended to illuminate.

At the end of the first millionth of a second in the history of the universe, it is thought that matter had cooled enough that quarks and gluons could not move around at will. Just like water freezes at 0°C (32°F), the quark-gluon plasma froze, leaving the quarks firmly ensconced in the resultant protons and neutrons. The universe would eventually cool further, allowing protons and neutrons to combine to make helium nuclei (consisting of two protons and two neutrons). Further cooling would let electrons attach to nuclei to make hydrogen and helium, which in turn would slowly coalesce into stars and galaxies.

But all the physical phenomena that govern that later cooling are relatively well known. It is the phase transition between quarks trapped in protons and neutrons and the free-ranging, low viscosity liquid quark-gluon plasma that heavy ion collisions at the LHC will explore. Perhaps the LHC will attain temperatures that could reach another phase transition from the low viscosity quark-gluon plasma to something more akin to a gas. Only time (and experiments) will tell.

Other Questions

The kinds of topics discussed thus far all tend to cluster near the frontier of knowledge and, for all of them, the LHC may reveal phenomena never before observed. However, in addition to frontier-blazing experiments, there are questions that scientists ask that are more evolutionary, that is to say the answers to these questions will improve and extend our understanding of phenomena about which we already know a great deal. In fact, all of the four large detectors at the LHC (ALICE, ATLAS, CMS, and LHCb, all described in chapter 4) will spend a great deal of their time on just these kinds of measurements.

However, there are two phenomena that will be studied at the LHC by small, dedicated experiments, called Totem and LHCf (for LHC forward). While the thrust of this book is the new horizons, the uncharted vistas that we hope the LHC will let us discover, these more progressive measurements are also interesting, and I mention them briefly here. These two phenomena are called *proton diffraction* and *cosmic rays*. I sketch both of them below.

Proton diffraction takes its name from an optical analogy. In optics, diffraction is the phenomenon whereby light waves can bend around corners. In fact, diffraction is a phenomenon exhibited by all waves, as shown in figure 2.14 in our discussion of quark structure searches. Recall that all particles have a wave equivalent, and the protons in the LHC are no exception. Thus it is expected that protons will exhibit diffractive behavior when they pass by one another.

Indeed, this kind of interaction between protons has been observed and studied for decades, most recently in detectors at Fermilab's Tevatron, in which protons and antiprotons are accelerated to an appreciable fraction (14%) of the LHC's energy. So studying this phenomenon at the LHC is expected to extend our current understanding, rather than open up an entirely new realm to study. Of course, there can always be surprises. This is the research frontier, after all.

The experimental signature for these kinds of studies is quite distinct. When the two protons collide, one or both of them survive the collision intact. One can contrast these kinds of collisions with the ones that are to be most fre-

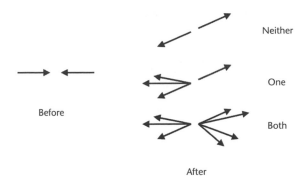

Before

After

Neither

One

Both

Figure 2.21. When protons collide, one, both, or neither protons might break up.

quently studied at the LHC, ones in which both protons are torn completely apart. Figure 2.21 illustrates the differences. While the destructive collisions are to be the most studied at the LHC, as they are the most violent and are most likely to reveal new physical phenomena, they are quite rare. Collisions in which at least one proton survives intact form the vast majority of proton-proton collisions. The Totem experiment is designed to explore this well-studied phenomenon at the higher energy the LHC will provide.

Cosmic Rays

Cosmic rays are a generic term for particles that rain down on the Earth from outer space. They were discovered well over a century ago, when physicists used the Eiffel Tower and the new-fangled hot air balloon to show that air was more conductive at great altitudes than it was on the ground.

Our understanding of cosmic rays has improved dramatically over the past hundred years. We now know that highly energetic particles from space, typically protons, hit the Earth's atmosphere and slam into an air molecule high above the Earth's surface, as illustrated in Figure 2.22. The most interesting cosmic rays are highly energetic, indeed much, much higher than any particle beam we can make on Earth. To give some perspective, the highest energy cosmic rays have about a hundred million times more energy than the beams at the LHC.

To study cosmic rays is actually rather tricky. Unlike in an experiment at a particle collider, you have very little information. You can't measure the energy or identity of the particle from space prior to the collision. You can't put instruments around the spot in the atmosphere where it collided with the air molecule. You can't observe exactly how the debris from outer space travels through the atmosphere. The only thing you can do is construct a detector that

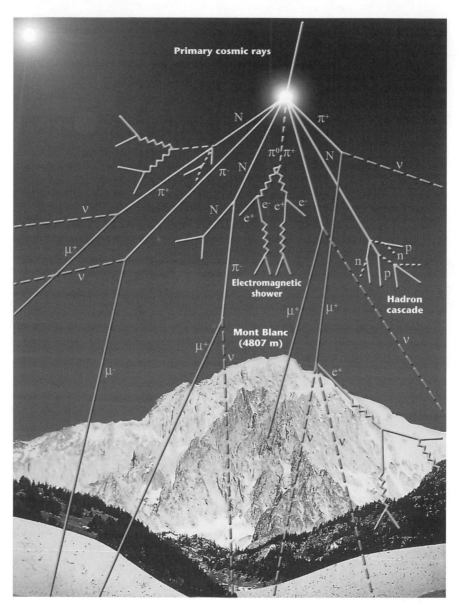

Figure 2.22. Protons from space hit air molecules, and the result is a shower of particles (shown by the letters *e* and *N* and Greek letters) that pass through the atmosphere and hit the Earth's surface. Courtesy CERN.

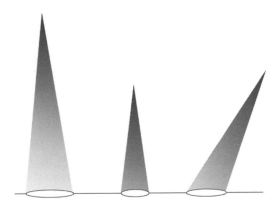

Figure 2.23. Three identical cosmic rays at different heights and angles of incidence. The relative darkness of the gray at the ellipse represents how much of the shower will be observed at ground level. This illustrates the need to know very well the height, angle, and general details of the cosmic ray shower's evolution.

sits more or less at sea level 20 to 30 km (10 to 20 miles) from the collision itself. From this meager information, we try to figure out the energy of the particle from space and even the direction from which the particle came.

Figure 2.23 shows what the cosmic ray experimenter is up against. Three identical cosmic rays interact in the atmosphere at different altitudes and with different incoming angles. The density of particles in the cosmic ray shower is most dense near the collision point and dies off as the particles are slowed and stopped in the atmosphere. In the figure, this loss of particles is illustrated by the lightening of the shower in the cone. The only measurement occurs in the ellipse at ground level. The figure clearly shows that these three identical cosmic ray showers will have very different signatures in the ground-based detectors.

To use the single measurement to determine the energy of the highest energy cosmic rays obviously requires a couple of things. First, you need a large detector, as a large shower can cover a vast area at ground level. Note that the Auger detector covers about 2,500 square km (1,000 square miles) of the Argentinean desert, or an area about the size of Rhode Island. The second is some method for determining both the direction and height of the start of the shower. This is usually done with timing techniques, but other methods are also used.

The third requirement and one most relevant to the LHC is a good understanding of just what happens when a very high energy proton hits another one. Without a good model of this process, the study is all guesswork. The problem is that there has never been a way to test our predictions of how such high energy protons interact with matter. The LHC will be able to provide tests never before possible. While not strictly speaking part of the LHC's main mission, the LHCf experiment will make measurements that will help out a lot.

In this chapter, we have only mentioned the tiniest fraction of questions that will be explored by the various detectors arrayed around the LHC. The Higgs boson, supersymmetry, and quark structure are some of the main topics that will be studied, but with about 5,000 experimental physicists involved, you can expect a veritable torrent of scientific results to come from these efforts.

What will be the next big discovery? I have no idea. It may well be one of the topics mentioned here. Or, even more exciting, it may be something utterly unexpected; something that just hits us out of the blue. As they say, time will tell.

While all these fascinating physics ideas are interesting to consider, it is only through experiment that we will know which idea is right. In the next two chapters, we will learn about the equipment, both the accelerator and the detectors, that will teach of this exciting and glorious frontier.

3

How We Do It

The Large Hadron Collider

Research is the process of going up alleys to see if they
are blind.

Marston Bates

Describing the pressing questions of modern physics is a noble
goal, but without an understanding of the equipment involved, it isn't possible
to appreciate the magnitude of the effort going into the search. This chapter
concentrates on the particle accelerator itself—the Large Hadron Collider—as
an instrument of discovery. The LHC is an extremely complicated device, 27 km
(17 miles) in circumference and comprising 1,232 primary magnets that take
6,900 km (4,300 miles) of wire to make. That's enough wire to stretch from New
York City to Las Vegas and back. But before we focus on the specifics of the LHC,
we need to spend some time on understanding the physical principles and technical ideas that go into the design of a modern particle accelerator. The first part
of this chapter discusses the main ideas involved in the design of any modern
accelerator and the end concentrates on the specific details of the LHC itself.

Fundamentally, the concept is this: we want to procure a source of protons
and accelerate them to the outrageous velocity of more than a billion kilometers
(670 million miles) per hour, which is about 99.999999% the speed of light.
Then we need to make a beam of these protons no wider than the width of a human hair and guide the beam in a circular path for a day or so (as shown in figure
3.1), during which time a proton will travel 26 billion km (or 16 billion miles).

All of these requirements sound pretty daunting, but the situation is made
worse by the need to have two such beams and to require them to hit exactly
head-on at specific times and places in the accelerator. Oh, and by the way, this
needs to be fairly simple to do, relatively quick, reliable, and at a manageable

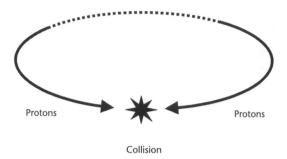

Figure 3.1. Modern accelerators send particles in a circular path moving in opposite directions. By making the beam particles (in the case of the LHC, protons) collide head-on, scientists create the most violent imaginable collisions and thereby study the most interesting physical processes accessible with that particular accelerator.

cost. I don't know about you, but the whole thing seems to be hard enough to verge on the impossible. And yet the CERN laboratory will accomplish just this. If you're not impressed by this, you've simply not understood the magnitude of this Herculean task. The Augean stables were simple by comparison.

Acceleration

So let's start with the basics. If you have a bunch of protons, how do you cause them to go fast? A slingshot? Draft a pitcher from Major League Baseball? Attach them to a three-year-old and feed them sugared breakfast cereal? Well, while all of these approaches might have their merits, the reality is a bit more practical.

If you have an object at rest and you'd like it to move, you need to use a force. We discussed in the first chapter the four forces, which are listed here ordered in strength from highest to lowest: strong, electromagnetic, weak, and gravity. Each of these forces affects different properties of matter. The strong force interacts with particles carrying the color charge. Since the proton has no net color, that rules out using the strong force. The weak force is, well, weak and only works over a short range, and so that's out too. The electromagnetic force interacts with particles carrying electric charge. Since the proton has electric charge, the electromagnetic force is a candidate. Gravity affects particles with mass, and protons are massive particles. Further, gravity is a long-ranged force and is therefore a candidate. However, as we noted in the first chapter, gravity is extremely weak, which means it is rather unsuitable for an accelerating force. This leaves electromagnetism and specifically electric fields to provide the impetus that will cause the proton to move.

It's actually pretty easy to understand how an electric field causes a proton to move. Fundamentally, it's a lot like gravity, with which we have ample fa-

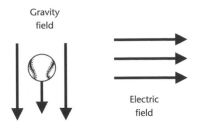

Figure 3.2. While gravity accelerates a baseball in a downward direction, an electric field accelerates a charged particle in any direction we desire. That's because we can easily change the orientation of the electric field.

miliarity. By way of analogy, hold a baseball up high and let it go, as shown in Figure 3.2. Gravity acts on the mass of the ball and it drops. Technically, we would call gravity a "gravity field" to indicate that the effects due to gravity are apparent over a large area.

Similarly, we can make an electric field that will interact with the electric charge of the proton and force it to move. We know how to make an electric field point in any direction we want and so naturally we orient the electric field so the proton is driven through our accelerator. Figure 3.2 illustrates these basic points.

So how does one make an electric field? There are lots of ways to do it, although not all are practical choices for a particle accelerator. But to give a basic idea of how it works, you can take a rubber balloon and rub it on your shirt. (A mylar balloon won't work.) Then take the balloon and run it just above your arm. You will find the electric field from the balloon tug your arm hairs. This experiment works best on a cool and dry day.

However, the simplest way to create an electric field that is useful for accelerator purposes is to take two plates of metal and connect them with wires to a battery, as Figure 3.3 illustrates. Between the two plates, an electric field will be set up, thus creating a simple particle accelerator.

Now when one makes such an electric field with a simple 1.5-volt, D-cell battery, the electric field isn't very strong and the proton isn't accelerated very much. Since we want to accelerate protons to extremely high speed, this is a technical problem that must be overcome. There are two solutions. The first is to simply use a stronger battery. This was the approach in old-style TVs (i.e., cathode ray tubes, on which most of my generation first watched *Gilligan's Island* or *Star Trek* reruns), which used a battery in excess of ten thousand volts.

Another approach is to take a bunch of particle accelerator units like those sketched in Figure 3.3 and stack them up. The particle would then be accelerated a little bit by the first one, more by the second one, and even more by the

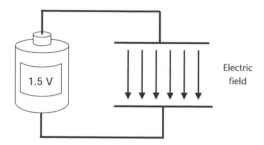

Figure 3.3. An electric field is simple to make with two parallel plates of metal, two wires, and a battery.

third one. If one wishes to make a gravitational analogy, if the accelerator apparatus illustrated in Figure 3.3 is equivalent to dropping a marble from a height of 1 meter, then two such sets of equipment is equivalent to dropping it from a height of 2 meters, and so on.

Modern particle accelerators use both techniques. Technology available when this book went to press in 2008 can make electric fields that are equivalent to attaching a 50-million volt battery to two metal plates separated by a meter (about 3.2 feet). However, remember that the point of this exercise is to use this technology to accelerate protons to a desired energy. Consequently, to understand accelerators, we need to understand how voltage and energy are related. Particle accelerators use a very convenient unit of energy. Rather than the familiar unit of kilowatt-hours found on your electric bill or even joules found in your high school science class, particle accelerators use measurements called *electron volts* (or eV, note both letters are pronounced). One electron volt is the energy an electron (or proton) gets when it is accelerated by a one-volt battery. Thus electrons in an old-style TV, with its approximately 10,000-volt battery, are accelerated to an energy of 10,000 eV. The LHC, with its ultimate beam energy of 7 × 10^{12} eV (or 7,000,000,000,000 eV) would require the equivalent of 7 × 10^{12} volts of batteries. This allows us to estimate how long a simple accelerator might be. For instance, with the accelerating voltage mentioned above (50 million volts per meter), would require more than 140 km (more than 84 miles) of batteries to achieve. Technically, you would accomplish this by taking 140,000 of these "two metal plates and a battery" contraptions and laying them end to end.

Good Vibes

So far, everything I've told you about how an electric field can be used to accelerate a particle is true, relevant, and I hope interesting. There's just a tiny little thing I forgot to mention. Electric fields in accelerators aren't made by the "two metal plates and a battery" idea. It's not that such an approach couldn't be made to work. The reason it's not used is because there turns out to be an easier way to make strong electric fields. Understanding how requires a little intellectual detour to the rural portion of America's rich heritage and to my Uncle Eddy's country jug band.

For those of you not lucky enough to have grown up in the deep country, a country jug band is a traditional and time-honored way to make country-style music. Such a band consists of various homemade instruments, including a washboard, a Jew's harp, and a contraption consisting of a metal washtub, broom handle, and twine. The most important member of the band for our purposes is the guy who blows across the top of a big moonshine jug. If he blows just right, the jug will emit a baritone sound that I've decided sounds like "huv." If the guy blows too fast or too slow, the jug doesn't make a loud sound. However at the perfect "blow speed," the jug emits a loud sound. The sound emitted by the jug is much, much louder than the sound of the guy blowing. You can reproduce this phenomenon yourself by blowing across the top of an empty 2-liter soda bottle.

The jug in jug bands is relevant to particle accelerators because it turns out that one can make the equivalent of a jug for electric fields. If you make a hollow metal container of the proper shape (the technical term is a cavity) and "blow on it" with radio waves, you can make very strong electric fields. Just how this works is a bit complicated, but is essentially the same phenomenon as the jug in Uncle Eddy's band. Because the way you "blow" on the cavity is using radio waves, we call this technique RF (for radio frequency).

Designing the precise shape of a cavity to make a strong electrical field is a highly complex task. Obviously, scientists want to make the strongest electric field in the smallest volume—to get the most bang for their buck, so to speak. Just as race car mechanics will use every trick at their disposal to increase the top speed of their car by a single kilometer per hour or to get a better cornering speed, accelerator designers exploit every technical trick in their book to make the most perfectly designed cavity to get the strongest electric field.

Luckily, we don't need to know all of these subtleties. Some of these cavities are shaped like a big bagel (Figure 3.4). Radio waves are aimed into the side of the bagel and the strong electric fields (the loud spot in the jug analogy) is at the hole of the bagel. This works out beautifully because we can aim our particles through the center of the bagel where the electric field is very strong.

This little detour into exactly how one makes electric fields seems a bit out of place, as this is a technical detail that one would ordinarily gloss over in a book of this type. However, we needed to take this detour, because it brings us to a point that we will need to comprehend just what is going on in the LHC.

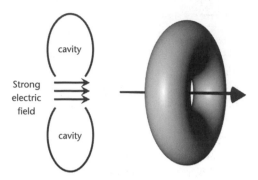

Figure 3.4. Radio frequency accelerators are elegantly simple structures. Here we show (*left*) a two-dimensional rendition of the basic geometry of an accelerating cavity, in which the strong electric field occurs in the center of a bagel-like structure and (*right*) a three-dimensional variant of the same. 3D variant courtesy Barry Panas.

Sound is a vibration. If you get near a pipe organ (which is essentially a very powerful version of our modest jug), you can actually feel the vibration when a bass note is played. Similarly, the radio waves that are used to excite the accelerator cavities are also vibrations (hence the "F" in RF). So this gets us back to the point of the detour. The electric fields in a particle accelerator are vibrating.

Note that this is somewhat different from the electric field and gravity analogy we drew a little while ago. After all, gravity is constant. But the electric field in a particle accelerator is constantly changing. It has a maximum, drops to zero, and even reverses direction entirely (although we'll ignore that particular fact in the following discussion, since it just unnecessarily complicates things). The upshot of all this is that we need to carefully time our beam's passage through the electric fields to occur when the field is strongest. That's the time when the charged particle will be accelerated the most. A consequence of this is that the charged particle will only pass through the electric fields at particular times.

We can most easily envision how this works by watching surfers (Figure 3.5). If you watch a surfer as she's starting to move quickly, you'll note that she's always in a particular spot on the wave. She's in front of the wave, about halfway up, somewhat closer to the top. That's the sweet spot, where she'll get the best ride. If the surfer were on the back side of the wave, she wouldn't move, nor would she move if she were in a trough.

This observation has interesting consequences. Suppose you had a bunch of very efficient surfers who wanted to take advantage of every wave. As each wave went by, each surfer would catch a wave. Thus, while surfing, the surfers

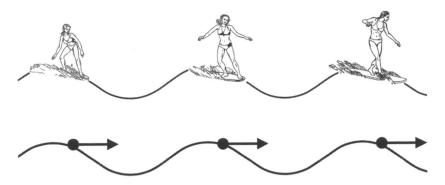

Figure 3.5. The use of waves by radio frequency electric fields imposes a structure on what a particle beam can look like. Just as surfers (*top*) can successfully occupy only a certain part of the wave, particles (*bottom*) can be accelerated only by certain parts of the oscillating electric field. Surfers courtesy Dan Claes.

would not get closer than a wave-separation apart. Similarly, when accelerating particles, the particles in one wave are always separated from ones in the adjacent wave. The separation depends on a particular accelerator's design, but in the LHC, particles riding adjacent waves are separated by about 7.8 m (25 feet). Note that particles don't *have* to ride each wave (for instance, in my own laboratory's accelerator, particles are only put in every 20 waves or so), it's just that they can't be closer than the separation of adjacent waves. Figure 3.5 shows the spaced structure of both surfers and particles.

Now we return from our detour and get back to the subject of accelerator length. I have already showed that the accelerator length of the LHC using the technologies discussed thus far could be over 140 km (or over 84 miles) long. Making a particle accelerator that is this long is not without its challenges. The most pragmatic challenge is cost. Basically, if you double the length of your accelerator, you double the cost. So any tricks you can play to shorten an accelerator's length are worth doing (as long as you don't have to pay some other expensive technical penalty).

Round and Round We Go

In the 1940s, scientists had a good idea about how to shorten the length of an accelerator needed. What if you could somehow use the same accelerator over and over again? Basically, we're talking about something akin to magic. Suppose you could drop a ball from a height of 3 m (10 feet) and when the ball got to the bottom you could use magic to make it disappear from the bottom (poof!) and reappear at the top (but with the same velocity it had at the bottom). If

Figure 3.6. A tetherball illustrates some of the important points of a synchrotron. A stationary person hits the ball (*left*), accelerating it. The rope guides the ball back to the person for another hit (*right*). This is entirely analogous to the single-point accelerating electric field and circular-motion-inducing magnets in a synchrotron. Courtesy Dan Claes.

you could repeat this action 10 times, then you could accelerate a ball with a 3-m-tall tower to a velocity that would ordinarily take a 30-m-tall tower without using magic.

Well, using magic in a physics laboratory is usually considered a poor choice, what with the rabbit fur and dove feathers getting into everything. No, the wizardry used to solve this particular problem was of a technical and not sleight-of-hand variety. This engineering magic is called a *synchrotron* (although there were predecessor technologies developed in the 1930s that used some of the same tricks). The principle that governs a synchrotron is essentially identical to that governing a tetherball.

A tetherball is a ball attached to a rope (Figure 3.6). The other end of the rope is attached to the top of a tall pole, anchored deep into the ground. A person hits the ball, and the rope makes the ball travel in a circular path. The ball comes full circle, and the person can hit it again. The ball goes faster and makes another circuit. If the rope is connected to the top of the pole in such a way that it doesn't wrap around the pole, in principle you can get the ball going very fast by synchronizing (and hence the name) both the orbit of the ball and the person's hitting it.

In a particle synchrotron, the electric field "hits" the proton and accelerates it. However, the counterpart of the rope in the tetherball analogy is not provided by electric fields but rather magnetic ones. Electric and magnetic fields are two facets of the underlying phenomenon called electromagnetism. While

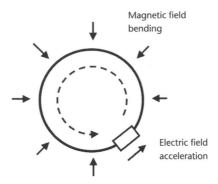

Figure 3.7. Schematic of a synchrotron. An electric field accelerates the charged particle and magnets guide the particle in a circular path (denoted by the solid line) back to the electric field, where additional acceleration occurs. The dashed line shows the sense of the particle's motion. In the case of the LHC, there are essentially two accelerators, each making a beam circulate in opposite directions.

we noted earlier in chapter 1 that electricity and magnetism are really the same thing, in an engineering venue we treat them differently.

Magnetic fields are handy in a synchrotron context because charged particles, such as protons, travel in a circular path when moving in a magnetic field. Figure 3.7 is a simple schematic of a modern particle synchrotron. Particles are accelerated by an electric field over a short distance and are then guided by magnetic fields in a circular path back to the electric field region for another round of acceleration.

Electromagnets

When I say that magnets are used to make particles orbit inside our accelerators, people often have different mental pictures of the magnets involved. Some think about the equivalent of the magnets that keep children's art on the front of the refrigerator. Others think of the horseshoe magnets that they played with in science class in elementary school. While both of these techniques create magnetic fields, particle accelerators use a more industrial approach. They need an electromagnet, which gets its name from the electricity used to make the magnetic field.

At its simplest, one makes an electromagnet by wrapping wire many times around a chunk of iron (Figure 3.8). Then you connect the wire to a battery. The electric current flows through the wire and makes a magnetic field. The iron isn't strictly necessary, but it acts as a type of amplifier, which greatly increases the strength of the magnet. You can easily make an electromagnet at home by taking a handful of iron nails and taping them together into a bundle.

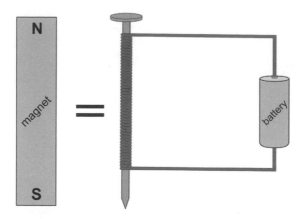

Figure 3.8. A simple coil of wire, wrapped around a nail and attached to a battery can make an electromagnet, which is a magnet created by electricity. The nail is actually not critical, but greatly strengthens the magnet. Courtesy Barry Panas.

Then wrap the bundle of nails with wire. The more loops the better, but 20 or so should do. Connect the ends of the wire to a battery, and you should be able to pick up other nails, paper clips, or similar metal objects. A scaled-up version of this simple demonstration is what is used in junkyards to pick up cars and move them around.

The magnets in particle accelerators are very similar in principle, although the engineering is a bit more precise. However, a coil of wire, a chunk of iron, and a battery are present in most accelerator electromagnets.

So why use the electromagnet technique? The first reason is the fact that magnets of this type can be very strong. Second, since you can shape the iron and wrap the wires in any way you choose, you can have great control over your magnet, choosing where the magnetic force must be strong and where it can be weak. The third and very important reason to use electromagnets is the presence of the battery. By changing the strength of the battery, you can alter the electric current in the wire, which in turn can vary the strength of the magnet. You can turn the magnet off, run it at full strength, or anywhere in between.

This brings us to an important point. The ultimate source of the magnetic field is the electric current passing through the coil of wire. More current means a stronger magnetic field. It's that simple. However, in general, most materials do not let current pass through them unimpeded. Materials resist the flow of current, and different materials resist the flow of current differently. Even copper, used in the wires in your house precisely because of its low resistance to the flow of electric current, does not let current freely flow through it.

While copper is a good material to be used to make electromagnets, it would be even better if new materials could be found that would resist the flow of electrical current even less. It turns out that there is a way to reduce a material's resistance to electrical current and to eventually eliminate resistance altogether.

It Doesn't Get Much Cooler Than This

In 1911, Dutch physicist Heike Kamerlingh Onnes was studying the electrical properties of materials as they were cooled. It had long been known that cooling a material reduced that material's resistance to the flow of electrical current. However, Onnes was a refrigerator specialist, and nobody could cool like he could. He was the first person to cool helium enough to turn it into a liquid. This is the same helium that fills children's balloons. Helium is the coldest liquid ever discovered and turns from a gas into a liquid at −269°C (−452°F).

For his electrical experiments, Onnes was cooling mercury and watching its electrical behavior, which was acting normally. Specifically, the electrical resistance of mercury was predictably dropping as the temperature was lowered. However, when the mercury's temperature got to −269°C (−452°F), something unexpected occurred. Suddenly it no longer resisted the flow of electrical current at all. A new phenomenon called superconductivity had been discovered. Rather quickly, many other materials were shown to be superconducting at similarly low temperatures. That the temperature at which helium liquefied and mercury became superconducting is the same has no particular significance. Other materials become superconducting at different temperatures.

Superconductivity was considered an interesting effect for decades (and explained in 1957), but became relevant to particle accelerators when physicists needed to make the strongest magnets possible. Strong electromagnets mean a lot of electrical current must flow in them. And, of course, creating a lot of electrical current is easiest when the electrical resistance is lowest. So naturally they decided to see if they could make large electromagnets from wires made of superconducting material.

The technical challenges were formidable, but eventually they were solved. In 1983, the Fermilab Tevatron, the first large accelerator made with magnets using superconducting technology was commissioned. Let's consider for a moment just how impressive a feat it is to have designed and built such a magnet. Bending magnets are called *dipoles* for reasons that will be explained shortly when we describe other types of magnets that do things other than bending the beam. Because this book is about the LHC, we would ordinarily use an LHC magnet as an example. However, the LHC magnets are a little trickier than most, so we'll show a more typical one at first.

Figure 3.9 shows a two-dimensional cross-section of a dipole that shows most of the essential features. The center of the dipole is a pipe through which

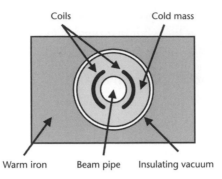

Coils   Cold mass

Warm iron   Beam pipe   Insulating vacuum

Figure 3.9. A cross-section of a dipole magnet. The center circle provides a path in which charged particles orbit. This is surrounded by a mass that holds the wire coils in place and keeps them cold. These are insulated from the outside world by a vacuum and surrounded by iron, which helps shape the magnetic field as desired. In this figure, both the particle beam and the current in the coils flow in the direction into or out of the page. Drawing courtesy Barry Panas.

the beam passes. The air in this pipe is pumped out, so the beam will not hit air molecules. The pipe is surrounded by tubes through which liquid helium can pass. On both sides of the hole are the coils of wire that make the magnetic field in the beam pipe. This is followed by strong structural material that keeps the coils from moving and then a series of cooling pipes, carrying ever-warmer material as one moves outwards from the center of the magnet. The bulk of the magnet consists of an iron support that gives structure to the whole magnet and holds it together.

Figure 3.10 shows a cross-section of the most common LHC magnet. The most striking difference from the ordinary magnet (seen in Figure 3.9) is that the LHC magnet has a passing resemblance to twins. There are *two* magnets buried in the surrounding metal and cooling equipment. This LHC bending magnet is about 14 m (45 feet) long and weighs 35 tons. The accelerator needs 1,232 of them. The coils at the center of the magnet consist of miles of wire and have to be cooled to about −271°C (−456°F), but the outer surface of the magnet needs to be room temperature. That means the center of the magnets need to be surrounded by what is effectively a high-tech thermos bottle. Even just sketching the basics of such a magnet reveals a daunting technological challenge, but of course a real-world magnet is even trickier.

Particle Beams

So far, we've discussed the challenging task of accelerating particles to near the speed of light and causing them to orbit in a circular path. However, let's not

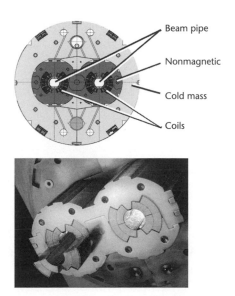

Figure 3.10. An example of an LHC dipole, which is more complex than the one shown in figure 3.9, as there are two dipoles, one for each beam. *Top,* a basic schematic; *bottom,* a photo of the center of a real LHC dipole. Courtesy CERN.

lose sight of why we have undertaken such a challenge. We want to collide a beam of subatomic particles head-on with another such beam. So let's examine what a beam of particles really means.

To begin with, a beam of particles requires more than one particle. The numbers vary depending on the accelerator, but having ten trillion protons in your particle accelerator at one time is not at all unusual. And, of course, when you have more than one particle involved, the situation gets complicated. By way of example, imagine driving the freeways of Los Angeles when you're the only car on the road. Now think about the traffic on a Friday afternoon at rush hour and you can easily see that the situation is far messier when you have more than one car (or particle!) involved.

Shortly we'll consider how the simple electric and magnetic fields mentioned earlier in the chapter must be modified to account for the real-world multiparticle beam. But, let's first discuss the basic geometry of a particle beam. Most people, when asked to describe the beam in an accelerator such as the LHC, will say "Huh?" and give you a blank look. But even among people who have pondered such things, they'll usually describe something like a stream of water coming out of a high-pressure hose. This picture is right in many ways, in that the beam starts at a point and travels in a straight line essentially forever. To

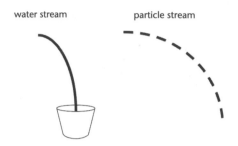

water stream   particle stream

Figure 3.11. A stream of water (*left*) is continuous as it flows from a hose into a bucket. In contrast, a particle beam (*right*) is discontinuous, with clumps of particles that are separated from each other. This discontinuous nature is caused by the manner in which particles are accelerated.

imagine this, think of a laser pointer. If you point it at your hand, you see a dot of light about a few millimeters wide. Now direct the beam at a wall across the room and you still see a spot that is few millimeters wide.

However, particle beams are not continuous. To see what we mean by continuous, imagine the water hose analogy mentioned above. If you could somehow instantaneously freeze the water coming out of the hose, you would see that there was a continuous stream of water from beginning to end.

In the particle beam case, things are different because there are gaps in the stream. If you could "freeze" a particle beam, you'd see a clump of particles, then a big gap, then another clump, and so on. The reason the beam has this feast-or-famine structure has to do with technical details of the accelerating electric field, which was mentioned earlier in our discussion of surfers.

The basic structure of a particle beam is the following: You have many clumps, or bunches, of particles. Each bunch is about 0.3 m (a foot) long and smaller than a millimeter (about a thousandth of an inch) in diameter. Each bunch is separated from its neighbors by several meters or even hundreds of meters (tens or hundreds of feet). Further, each bunch can contain hundreds of billions of protons. In Figure 3.11 I try to convey the essentials of any particle beam. The specific numbers vary for any particular accelerator. For the LHC, there will be 2,808 distinct bunches, initially with 100 billion protons per bunch. These bunches will be separated by no less than 7.8 m (25 feet).

There are three important parameters in discussing any accelerator: the type of particles being accelerated, the energy of the beams, and the number of collisions per second. More collisions per second are good. In an accelerator like the LHC, two beams of particles are aimed at one another. In many respects, it's like two rifles shooting bullets at one another, with the hope that the bullets will collide head-on. In the case of the LHC, two bunches, each containing hundreds of billions of protons, are made to pass through one another, with the

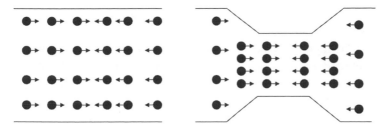

Figure 3.12. The particles in the respective beams can collide when particle beams pass through one another. However, if the beams can be focused, so the particles in each beam are closer to one another, the probability of collisions is increased.

hope that maybe two protons will collide with sufficient violence to do something interesting.

Focusing on the Problem

So how would we increase the probability that two protons will get close enough together to collide violently? The trick is to make the beam width as tiny as possible. To see what I mean, let's imagine two swarms of bees flying through one another. If you had two such swarms and wanted to make a head-on collision between two bees more likely, what would you do? You'd force the bees within each swarm to fly closer to their neighbors.

In Figure 3.12, we see the effect of forcing the bees in a swarm to fly in tighter formation. Because there is less space between adjacent bees, as the swarms pass through one another, the probability of a head-on collision increases. Of course, we are interested in a particle accelerator and not a bee accelerator, but the basic idea holds. To increase the probability of violent collisions between protons, and thus that the accelerator will provide us with valuable information, one can do two things. First, put as many protons in your accelerator as possible and, second, make your beams as narrow as possible.

The language of this aspect of particle collisions borrows heavily from light and lenses. We talk about the luminosity of a beam, which is analogous to the brightness in a beam of light. A beam of light can be made brighter by increasing the amount of light in it or by using a lens to focus it.

So this naturally leads us to ask the questions, "How does one make a particle lens?" or, more generally, "How does one focus a particle beam?" To answer this question requires that we return to our discussion of magnets. If you recall, the magnets that forced the particles to go in a circular path were electromagnets, consisting of coils of wire and chunks of iron. Each bending magnet consisted of one coil of wire. These lenslike, or focusing, magnets use two coils. The technical names for these magnets are *dipole* (for the one-coil, bending magnets) and *quadrupole* (for the two-coil, focusing magnets). However, we will continue to simply call them bending and focusing magnets.

While the superconducting bending magnets are the biggest and strongest types of magnet in any circular particle accelerator, the focusing magnets are no less crucial (indeed most accelerators also contain three-, four-, and five-coil magnets as well). Focusing magnets are critical because many factors conspire to make the beam much wider than desired. The simplest are the manufacturing and installation imperfections in the magnets that cause them to operate less than perfectly.

Another big concern arises from the self-destructive nature of the beam itself. Recall that the LHC beam consists of vast numbers of protons, which are positively charged particles. You may have heard that opposites attract, and that may work in dating, but it certainly works in electricity. In electricity, objects with opposite signs attract one another, while objects with the same-sign electric charges repel one another. Consequently, the electrical self-repulsion among the protons in the beam itself causes the beam to grow much larger than its collision-optimized small size. The focusing magnets are constantly squishing the beam, much like a magnifying glass will focus sunlight.

To give a sense of scale, we can mention the raw numbers of the different types of magnets in the LHC. There are 1,232 bending magnets, spread uniformly over the 27-km (17-mile) circumference. In addition, there are 858 focusing magnets. There are also 7,210 smaller "correction" magnets, for a grand total of 9,300 magnets. These more complex magnets are used to correct for the fact that not all protons in the accelerator have precisely the same energy, as well as other small effects. These three kinds of magnets work together like a team of kindergarten teachers trying to get a group of kids moving as desired: "Go left! Do it again! Bunch up and stay together! Hey! Back in line! Yes, I'm talking to you!"

Thus far, we've discussed the LHC as if it were a single circular accelerator, and that's true in a very strict sense. However, it's also misleading. It's more accurate to say that the CERN laboratory hosts an entire bevy of other accelerators, each crucial for the LHC's mission. The need for a series of accelerators is true for any modern laboratory. The logic is no different from what goes on in any automobile. Essentially, each accelerator can be treated as a different gear in a car. If you've ever driven a car with a manual transmission, you know that it is possible to take a stopped car and get it moving only using the highest gear. However, this is very hard to do, and you run the risk of frequent stalls. It's simply easier and more efficient to shift the car through a series of gears, each carefully tuned to match engine and wheel speed.

So, too, it is with accelerators. In the LHC, one has to accelerate stationary protons to nearly the speed of light (300,000 km, or 186,000 miles, per second). In principle, you could build a single accelerator that could span that entire range of energy, but the fact is that technical, engineering, time, and cost concerns make that a very poor choice. It is simply more efficient to have a series of accelerators, each tuned to a different energy. For example, the first might accelerate a proton from rest to 1% of the speed of light, and the second might go from 1% to 10%. The third might go from 10% to 80%, and so on. Such a "chain" of accelerators is what makes up the LHC complex, with the actual Large Hadron Collider being only the highest energy accelerator in the chain. However, without all of the steps in the chain, the LHC would be nothing more than a very expensive tunnel, connecting Switzerland and France.

A Brief History of CERN

One doesn't need to be much of a history buff to know that the early 1940s was a horrible time for Europe. The jackbooted shadow of storm troopers darkened most of the continent until 1944, when the largest amphibian invasion in history led to the desperate hedgerow fighting of Normandy. An endless series of trains rattled eastward toward camps, carrying their grim and tragic cargo and searing the camp's names into our collective psyche. Wave after wave of British bombers rained down fire on Hamburg, creating such a maelstrom that the air itself seemed to burn, reducing the city to ashes and killing 43,000 souls overnight. Fighters above Britain banked and rolled violently in a dance of death. The Red Army reduced Berlin to rubble. Europe was convulsed in a war so terrible that the destruction and suffering has not been matched before or since.

This consuming conflict ended in May 1945, after which the victorious Allies conducted the Nuremburg trials, accusing, convicting, and condemning to death the Axis leaders for crimes against humanity—crimes so heinous that they needed a new name. The continent had been torn asunder.

Thus it remains astonishing to me that just a mere four years later, in 1949, Nobel Prize–winning physicist Louis de Broglie could propose a new pan-European physics laboratory in which all of the major European powers (victors and defeated alike) would participate. Just three years after that, 11 European governments agreed to create the provisional Conseil Européen pour la Recherche Nucléaire (European Nuclear Research Council), or CERN. With the ratification by the member nations of the treaty setting up the organization on September 29, 1954, the provisional status disappeared and the current name of European Organization for Nuclear Research came into being, although the CERN acronym was retained.

The signatories to the treaty were Belgium, Denmark, France, the Federal Republic of Germany (or West Germany), Greece, Italy, the Netherlands, Norway, Sweden, Switzerland, the United Kingdom, and Yugoslavia, one more than the original 11 nations. Less than a decade after the bombs stopped falling, at least the scientific European community was healing. Perhaps even most surprising was the fact that the voting public in the signatory nations went along.

In 1952, Geneva, Switzerland, was chosen as the site for the new physics laboratory and, in 1957, CERN's first particle accelerator came online: a 600-million-eV synchrocyclotron. This initial accelerator operated with about ten thousand times lower energy than today's LHC.

An accelerator of that energy was noteworthy more for its mere existence, initiating as it did CERN's 50-year presence (so far) as a juggernaut on the world's stage of particle accelerator laboratories. But, in 1959 CERN catapulted itself to the top of the particle physics world, when it turned on the 28-billion-eV Proton Synchrotron (PS), briefly the world's highest energy accelerator and still operating today as part of the LHC complex.

These early accelerators all had a common feature. They all worked in what is known as *fixed target mode,* which simply means that experiments were all done by shooting the beam into a stationary (i.e., fixed) target. This is akin to someone shooting at a wall or, given that we are discussing circular accelerators, akin to the sling with which David is reported to have slain Goliath. For those of you for whom biblical stories are a bit rusty, a sling is used to spin a rock in a circular motion. After the rock is moving at great speed, it is released and travels in a straight line toward its target. In the case of David and Goliath, the target was Goliath's forehead.

In fixed target experiments, a particle beam is accelerated via the methods we've discussed earlier in this chapter, and the charged particles are then aimed at a target. The target can be anything, although it is typically a container of hydrogen, chilled until the hydrogen liquefies.

However, the LHC does not operate in fixed target mode. The C in LHC stands for "collider," after all. In such a configuration, two beams of counter-rotating beams of charged particles are aimed at one another to collide head-on.

What are the advantages of colliding beams over fixed target experiments? The disadvantages are obvious. Aiming two beams at one another is hard, while in the fixed target case, you only have to aim your beam at a target, which you can make arbitrarily large. The phrase "hitting the broad side of a barn" comes to mind. So why a collider?

The reason is energy. When a projectile hits a stationary target, the debris from the collision ends up moving in the direction of the initial projectile. For a visual image, think of shooting at a watermelon. When you hit the watermelon,

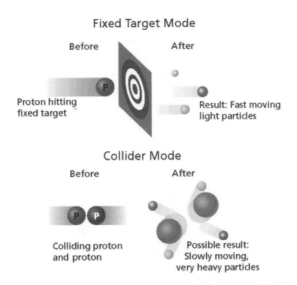

Figure 3.13. When a beam particle is aimed at a stationary target (*top*), the laws of physics do not allow a very violent collision, as much of the energy goes into moving the debris. In contrast, when a particle collision occurs head-on between two objects with identical energy (*bottom*), the two particles can stop dead in their tracks and all the energy can go into creating new physical phenomena. Courtesy the Particle Data Group, Lawrence Berkeley Laboratory

most of the pieces will blow out the back. However, in the collider configuration, the two projectiles can simply stop, like two cars involved in a head on collision. When this happens, the result is that all of the energy of the collision would go into doing something "interesting." In our analogy, the interesting thing would be to thoroughly wreck the cars. In our particle physics world, "interesting" simply means revealing some new and rare physical phenomenon.

The difference in the available energy is surprisingly large. The LHC has two beams, each carrying 7 trillion electron volts of energy. When particles in the opposing beams hit head-on, there are 14 trillion electron volts of energy available to possibly discover something new. However, if just one of these beams were to hit a stationary proton, the useful energy would not be the full 7 trillion electron volts of energy. Rather, the energy available would be a meager 0.1 trillion electron volts, or less than 1% of the energy in the head-on case. Figure 3.13 illustrates the basic differences between fixed target and collider operations.

The LHC is the latest iteration of a type of proton-proton collider first proposed in 1965. Commissioned in 1971, the ISR (for Intersecting Storage Rings) had an available collision energy of 62 billion electron volts. The ISR didn't have detectors of the type to be found at the LHC, so it was much more of a triumph in accelerator design than in physics. For instance, it held the world record for the brightness of the beams at the collision points until 2004, when the United States' Fermilab Tevatron finally surpassed it, fully 20 years after the ISR's shutdown in 1984.

The year 1971 included a proposal for a new accelerator at CERN, the 6.3-km (3.8-mile) circumference SPS (Super Proton Synchrotron). The SPS began operation in 1976. Its design energy was 300 billion (3×10^{11}) electron volts, although CERN's clever accelerator scientists were eventually able to run it at 500 billion electron volts of energy, or 7% the energy of today's LHC. The SPS project came in ahead of schedule and under budget and still plays an important role in the LHC complex.

The SPS now operates as one of the accelerators in the LHC chain; however, it briefly was reconfigured to run in a bold, new way. The SPS was turned into a colliding ring. This was not so innovative; after all, the ISR had already done that. No, the innovation was that these colliding beams would consist of protons—still nothing new there—and—the innovation—*antimatter* protons! While America's Fermilab Tevatron is currently the world's best proton-antiproton collider, it wasn't the first, nor is it at all clear that the Fermilab accelerator would have turned on as cleanly as it did without the Sp$\bar{p}$S (Super Proton-Antiproton Synchrotron) leading the way.

The next noteworthy date in our whirlwind trip through CERN's history is 1981, when the 27-km (17-mile) circumference LEP (Large Electron Positron) was approved. This accelerator was eventually designed to accelerate electrons and antimatter electrons (positrons) to an energy that was precisely selected to produce enormous numbers of Z particles. You may recall from chapter 1 that the Z particle is a carrier of the weak force.

In 1989, the LEP accelerator turned on and performed simply brilliantly. The measurements performed by the four competing experiments on the properties of Z particles may never be surpassed. My personal favorite LEP measurement is the one that is generally interpreted as proving that there are three and only three particle generations. (Remember that a generation is one of the carbon copies of subatomic particles, of which the up and down quark and the electron and electron neutrino are the first.) Technically, the measurement showed that there were only three light neutrinos, which leaves open the possibility of additional generations as long as they have heavy neutrinos. Either way, this measurement is an important clue to the particle puzzle. I just wish I could figure out what it is telling me. I say that a lot, but the universe is trying to tell us something profound. When someone figures it out, I'm going to say "Well duh! I wish I had thought of that!"

A little side note illustrates the caliber of the effort that went into the accelerator's construction and operation. To understand the data gleaned from the accelerator, the CERN scientists needed to set it to a precise collision energy. After a while, during which time they commissioned the accelerator and experiments, they saw something peculiar. The energy of the beam varied over the course of a day. There appeared to be two cycles per day, with the time between cycles being about 12.5 hours and the time of day during which the maximum and minimum deviation in energy occurred about an hour later each day. Very peculiar. Equipment was checked, heads were scratched and, um, colorful language was uttered. Eventually an innovative thinker had the long-sought "Aha!" moment. It turned out that the mysterious variation was caused by the effect of the lunar tides on the Earth. The force of the moon's gravity causes the surface of the Earth to flex by about 0.3 m (1 foot). This had the effect of changing the radius of the 27-km (17-mile) circumference LEP accelerator by about 1 millimeter. This works out to be a change of 0.00001% and it was noticeable. Wow.

The LEP experiments made hundreds of precise measurements and searched for new phenomena, including looking for the top quark before losing out that honor to Fermilab in 1995. Before the LEP accelerator was decommissioned in November 2000, it was run at even higher energies, eventually more than double that of its original design. The idea was that if you're going to turn off a piece of equipment soon, you have little to lose if you damage it in the attempt. I have a mental image of the CERN accelerator scientists calling up the control room and saying "Captain . . . the main energizer is bypassed like a Christmas tree . . . I canna guarantee how long she'll last. But you've got power" (with apologies to James Doohan and Gene Roddenberry). And yet deliver they did. It makes me wonder how many of those accelerator scientists were Scottish.

In its last few months, the LEP accelerator delivered data that seemed to indicate that perhaps its experiments had discovered the Higgs boson. While these data are no longer believed to have supported that idea, at least LEP went down swinging and the four experiments' results still play a very prominent role in our current understanding of the Higgs boson.

The LEP accelerator was decommissioned in 2000 for a reason very important to readers of this book. This reason is because the LHC now inhabits the LEP tunnel. Out with the old and in with the new, as they say, and a new era has begun.

Long before the LEP accelerator was decommissioned, the CERN governing council decided that putting the LHC in the LEP tunnel was the future of European particle physics. In December of 1991, this fateful decision signaled the eventual death knell of the LEP accelerator. The construction approval in 1994 sealed its fate. With the writing on the wall being rather apparent, the four LEP experiments put the remaining time to excellent use. There is no greater epitaph for any experiment.

The period of time between 2000 and 2005 was marked by feverish activity. The LEP accelerator and all four experiments needed to be dismantled and removed. The LHC components needed to be assembled, with the first of the bending magnets installed in 2005. Final magnet production occurred in 2006, and the final bending magnet was lowered into place on April 26, 2007.

With our quick journey through CERN's 50-year history now complete, we finally come to the point of this chapter: a discussion of the LHC accelerator complex.

Nuts and Bolts

CERN is a fairly small site on the Franco-Swiss border. Most of CERN's accelerators are contained within the site's perimeter. However, this is not true for the LHC itself. The LHC breaks free of the site and swoops in a large circle through the French and Swiss countryside. Well, to say it passes through the countryside is misleading. It actually passes under the countryside, on average 100 m (about 300 feet) underground. The depth varies from 45 to 175 m (150 to 560 feet), depending on the location's proximity to the foothills of the Jura Mountains. Thus the LHC accelerator is invisible to the surface dwellers. People living above the ring pass their lives in blissful ignorance of the frantic dance of protons circling under their feet. Figure 3.14 shows a bird's eye view of the area surrounding CERN, while Figure 3.15 shows more clearly the subterranean nature of the LHC ring.

The LHC complex consists of five distinct accelerators and some equipment preceding the first real accelerator. Prior to actual particle acceleration, one must obtain protons. One does that by taking ordinary hydrogen, which consists of a proton and electron and stripping off the electrons. This is done via the Duoplasmatron source, which seems to have stolen its name from 1930s science fiction. The Duoplasmatron source provides protons with an energy of a hundred thousand electron volts (1×10^5 eV). This meager energy works out to be about 1.5% the speed of light but is still about 2,700 miles per second.

The first real accelerator a proton encounters in its journey is Linac 2. A linac (short for linear accelerator) is a straight-line accelerator, consisting of electric fields all pushing in the same direction. A proton enters the 78-meter- (256-foot-) long linac with essentially no energy and leaves it with an energy of 50 million electron volts (5×10^7 eV). Fifty million of anything sounds like a lot, but it is far less than the LHC's ultimate energy of 7 trillion electron volts (7×10^{12} eV). A proton leaving the linac has about a millionth of the proton's final energy. Even so, the velocity of a proton leaving the linac is traveling 31% of the speed of light.

As the proton leaves the linac, it is guided into the first circular accelerator, the Proton Synchrotron Booster (PSB), a ring with a circumference of about 157

Figure 3.14. Satellite image showing the location of the LHC, just outside Geneva, Switzerland. The 27-km-long (17-mile-long) ring spans the Swiss and French border (shown with a dotted line). The Jura Mountains are in the upper left. Courtesy CERN.

meters (515 feet). The PSB increases the proton's energy from the linac's 50 million electron volts to 1.4 billion electron volts (1.4×10^9 eV). The proton's velocity leaving the PSB is 91.6% the speed of light.

The next accelerator is the PS. Once the world's highest energy accelerator, it is now but a way station on the way to the LHC. The PS is a ring with a 610-m (2,010-foot) circumference, and it raises the energy of the proton to 25 billion electron volts (2.5×10^{10} eV). The proton velocity leaving the PS is 99.93% the speed of light.

The penultimate accelerator is the SPS, once the site of the discovery of the W and Z particles (discussed in the first chapter). This fourth accelerator has

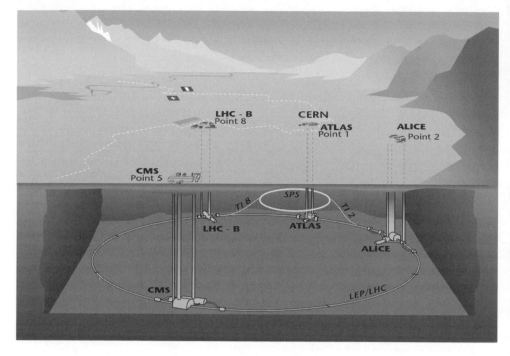

Figure 3.15. The LHC accelerator complex, with the relative locations of the various experiments denoted. Note that the vertical scale is misleading, as the detectors are about 90 m (300 feet) underground and the ring is 27 km (17 miles) around. There are eight points around the LHC at which you can place detectors, with points 1, 2, 5, and 8 currently occupied. TL indicates transfer lines between the Super Proton Synchrotron (SPS) and the LHC. Courtesy CERN.

a circumference of more than 6.3 km (3.8 miles) and raises the energy of the proton to 450 billion electron volts (4.5×10^{11} eV). These protons, now traveling at the astonishing speed of 99.9998% the speed of light, are now ready for injection into the LHC.

The fifth and final accelerator is the LHC itself. The LHC accepts the protons from the SPS and increases their energy to the full 7 trillion electron volts (7×10^{12} eV). This is 100% of the design energy and brings the proton's velocity to 99.9999991% the speed of light. This is so fast that a person sitting in one of CERN's laboratories will see a photon only moving 2.7 meters (8.8 feet) per second faster than the LHC's protons. You'll note that the velocity in this final accelerator is not much faster than it was in the PS, although the energy is about 300 times greater. This is a consequence of Einstein's theory of special relativity, which describes how things move at high velocities, but it still might seem odd to many readers.

In all of the circular accelerators except for one, the beams orbit either in

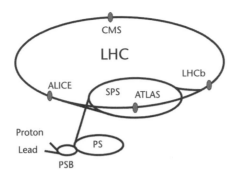

Figure 3.16. Schematic of the LHC accelerator complex, with all accelerators indicated by initials. The light ovals denote the four major experiments that were discussed in chapter 4. All five accelerators are listed: (1) Proton and lead indicate the first (linear) accelerators and the beams they carry, (2) PSB is the Proton-Synchrotron Booster, (3) PS is the Proton Synchrotron, (4) SPS is the Super Proton Synchrotron, and (5) LHC is the Large Hadron Collider itself.

a clockwise or counterclockwise manner. However, the LHC consists of two beams, one orbiting in each direction. In Figure 3.16, we see that there are two different points at which protons can be transferred from the SPS to the LHC, enabling a clockwise-only accelerator to load the LHC with protons in both directions. This is just one of many clever engineering features embodied in this intricate equipment.

Every one of the five accelerators has electric fields formed by the RF cavities discussed earlier in the chapter. Further, all of the accelerators except for Linac 2 have bending magnets. Rather than discussing all of the details of all of the accelerators, let's focus on the LHC.

To describe the LHC really involves a long list of interesting numbers. Each of the counter-rotating proton beams are accelerated by eight RF cavities. Each RF cavity adds 2 million electron volts of energy, thus giving each beam an additional 16 million electron volts of energy on each orbit. So, in the absence of other considerations, the proton beam could be fully accelerated in 400,000 orbits. Given that the beam makes a little over ten thousand orbits per second, the acceleration phase could take as few as 40 seconds.

The actually acceleration time is much longer—more like 20 minutes. The acceleration period is much longer in part because the electric current in the various magnets must be increased to strengthen the magnetic fields as the beam energy is increased. To keep the beams inside the beam pipe (which you may remember is a tube a few centimeters, or a couple of inches wide, and 27 km, or 17 miles, long), the magnetic fields must be constantly adjusted to tame the beams' increased energy.

The bending magnets are capable of running with a magnetic field of 8.3

tesla, which is about 170 thousand times more powerful than the Earth's magnetic field. The Earth's magnetic field is what makes a compass such a useful tool. This incredible magnetic field is achieved by running 11,800 amperes of current through the wire of the bending magnets. The power distribution panel for my house is rated for a maximum of 100 amperes, so each bending magnet will use the current necessary to power over 120 ordinary houses. The entire string of 1,232 bending magnets consumes the electric current that could supply the needs of a small city, consisting of 150,000 houses. When the entire LHC complex is considered, the total power consumption is about 120 million watts of electrical power at peak demand.

With so much electrical energy stored in the magnets, it is natural to wonder how much energy that works out to be. The total energy stored is an astounding 11 billion joules. To give some perspective, that is about the amount of energy stored in two and a half tons of TNT, although this energy is spread out over the whole 27 km (17 miles). This is equivalent to a 400-ton 747 jet airplane hitting the ground at about twice the speed of sound.

Each bending magnet is about 14 m (45 feet) long and weighs about 35 tons. It is necessary to cool the coils that carry the electric current to –271°C (–456°F). These cables are made of a superconducting niobium-titanium alloy and look somewhat like the kind of wire that powers household appliances. Cut the power cable to a table lamp, and you'll see that each cable is made of smaller strands, twisted like a rope. In the case of the LHC's superconducting wire, the wire is made of strands, which are in turn made of filaments. The length of strands that make up all the LHC's bending magnets would go around the Earth's equator about seven times. The length of the filament involved would stretch five times from the Earth to the sun and back, with enough left over for a few trips back and forth to the moon.

To cool these magnets requires an extraordinary amount of coolant. A total of 10,800 tons of liquid nitrogen is needed, followed by 120 tons of liquid helium. The entire cooling process takes about six weeks, during which time the 37,000 tons of magnets are cooled from room temperature to more than –271°C (–456°F).

Although it takes four and a half minutes to transfer enough protons from the SPS to the LHC and about 20 minutes to accelerate them to their maximum energy, the beams are then left to collide in the detectors for 10 to 20 hours. For the beam to last so long, the pipes through which the beams circulate must be under a superb vacuum. If a vacuum were not in the pipe, the beams would interact with the molecules of air in the pipe and immediately disappear. The vacuum in the beam pipe is ten trillion times rarer than ordinary air (10^{-13} atmospheres for the technical crowd). This is among the best vacuums achieved on Earth. Perhaps even more impressive is the total volume that needs to be a vacuum: 6,500 cubic meters (over 220,000 cubic feet), or about equivalent to pumping all of the air out of one of Europe's many majestic cathedrals.

The last topic to cover is the structure of the LHC beam itself. As mentioned before in our discussion of surfers, the beam is comprised of bunches of protons, with each bunch separated by no less than 7.6 m (25 feet). You may recall that this is the distance between adjacent waves in the accelerating electric field. Of course, it is possible for adjacent bunches to be farther apart than this. If not every wave is filled with protons, adjacent bunches could be multiples of this distance, depending on how many waves are skipped.

So let's look at an individual bunch. One bunch in the LHC includes about 100 billion (10^{11}) protons. The actual shape of each bunch has a passing resemblance to a stick of uncooked spaghetti, although it is about 0.3 m (1 foot) long and the width is less than a millimeter. There will be 2,808 bunches of protons orbiting in each direction and aimed and focused to collide at four points around the LHC's perimeter. So, except for the actual beam width being about a hundred times smaller than a piece of spaghetti, you can get a pretty good visualization of the LHC's beam as about 3,000 pieces of uncooked spaghetti, each separated by 7.6 m (25 feet), orbiting at 99.999999% the speed of light, in an orbit that is 27 km (17 miles) around. If you do the math, you find that about 3,800 lengths are needed till up the entire orbit. So if there are 2,808 bunches, the entire accelerator is not filled. You have a concentrated group of bunches, each separated by the 7.6 m (25 feet), followed by a relatively long gap. This gap has many uses. Note that "long" is a relative term and is in the ballpark of a millionth of a second. This exceedingly brief time during which there are no protons in a detector is used for the detector to recover and reset itself.

The energy stored in these beams is enormous, although only about 3% of that stored in the magnets. However, this beam must pass through the center of the various detectors in the center of the experiments spaced around the ring. The equipment in the experiments that is near the beam is extremely delicate and consequently extreme care is taken to be able to nearly instantaneously dump the entire beam into a large absorber if there is the slightest indication that the accelerator operators are losing control of the beam. The beam carries so much energy that were it not controlled so carefully, it could easily destroy the heart of the particle detectors. To give you an idea of how much energy we're talking about, it's about 350 million joules. That's as much energy as a 400-ton commuter train traveling at 160 km (100 miles) per hour, or enough to melt a half a ton of copper. To protect the equipment, this energy is diverted to a stack of graphite absorbers in under a thousandth of a second. That's like absorbing the energy of a fairly large military conventional (i.e., nonnuclear) air-dropped bomb.

These two counter-rotating beams will each circle the LHC accelerator ring a little over ten thousand times per second. Something like 800 million collisions per second will occur in each of the detectors, although most collisions will not be especially interesting. During their 10 to 20 hours of collision time, the beams will travel about 16 billion km (10 billion miles). That's about like traveling to Neptune and back, all the while circulating in a pipe a few centimeters (or inches) wide. No matter how you look at it, the LHC is an extraordinary technological marvel.

Lead Beams, Too

Before we close this chapter, we need to discuss one additional thing. We've focused predominantly on the case where the LHC is colliding beams of opposing protons. But the LHC is not a one-trick pony. The LHC is also designed to be able to accelerate heavy ions, which are atomic nuclei stripped of all their electrons.

Although the LHC can accelerate many different heavy ions, its design is optimized to accelerate lead nuclei. Lead consists of 82 protons and 126 neutrons. The process whereby lead is accelerated is similar to the proton case, so we'll only discuss the main differences.

A pure sample of lead is heated to about 538°C (1,000°F). An electric current is passed through the lead to knock some of the electrons off the lead nuclei. Each lead atom also contains 82 electrons. The current can typically knock about 30 electrons off, but usually not many more. The lead nuclei are accelerated through a different linac, called Linac 3, and the lead beam is passed through a thin carbon target, which knocks off another 20 electrons or so. To accumulate sufficient lead to make enough collisions of interest to the experimenters, the beam is guided into a storage accelerator called the Low Energy Ion Ring (LEIR).

When enough lead has been stored in the LEIR, the lead is then transferred to the PS accelerator, which accelerates the lead and passes it through another target, which knocks off the remaining 30 or so electrons. The lead nuclei, now stripped of all their electrons, are passed through the SPS into the LHC. In the LHC, the lead nuclei are accelerated to 2.8 trillion electron volts per nucleon. Recall that when the LHC is accelerating protons, each proton carries 7 trillion electron volts. So it first seems like the LHC lead beams are lower in energy than the proton beams. However, recall that each lead nucleus contains a total of 208 protons and neutrons. So the total beam energy per lead nucleus is about 575 trillion electron volts. This will result in the most violent, large-volume collisions ever recorded. The LHC will run in the mode in which heavy ions are accelerated about one month a year.

By any measure, the LHC accelerator is a highly complex instrument. It is intended to concentrate an unprecedented amount of energy into incredibly tiny volumes. However, no matter how impressive a technical achievement the LHC accelerator is, if the collisions are not recorded, the whole exercise is pointless. In the next chapter, we will discuss how these particle collisions can be recorded.

4

How We See It

The Enormous Detectors

The great tragedy of science is the slaying of a beautiful theory by an ugly fact.

Thomas Huxley

Countless particles travel in circles under the Swiss countryside, occasionally bumping into each other. All of the effort put into accelerating these particles is in vain if we do not record the collisions between protons by taking what amounts to fast and high-tech photographs. By recording and reviewing millions and indeed billions of these collisions, we will begin to understand what can happen in collisions between protons, from the common to the rare. Finally, by understanding why the common things are common and the rare things are rare, we will learn a great deal about the behavior of matter and energy under extreme conditions and even about the birth of the universe itself.

So just how does one record the collisions caused by a large particle accelerator such as the LHC? You need huge detection equipment, weighing thousands or tens of thousands of tons. In a collision likely to occur in the LHC, two particles enter the collision (the protons) and lots (say somewhere in the neighborhood of 10–500) of particles come out. The story of each collision is etched in the trajectories and the identities of the outgoing particles. Because the actual collision occurs in such a mind-bogglingly short time, the collision itself is usually hidden from us. It's only by looking at the debris of the collision that we can observe what we need to answer our questions. Understanding particle collisions is essentially a study in forensics.

You can understand the mind-set of an experimental physicist if you pretend to be a bomb-squad investigator. Bomb investigators do not generally understand the details of the explosion by being close by when it occurs—at least not if they want a second assignment. No, bomb investigators understand the explosion by studying its effects on its surroundings. By studying scorch marks, total damage, amount of debris, and how deeply the shrapnel penetrates well-understood materials, the expert can get a good idea as to what happened. Chemical analysis adds to the story.

Similarly, particle physicists study their collisions by surrounding the collision point with a detector of well-known and carefully selected composition. By seeing how the particles leaving the collision interact with the detector, their energy, trajectory, and point of origin can be inferred. The right detector will reveal at least some of the particles' identities. This information can be brought together like a jigsaw puzzle, with each piece of information neatly interlocking and revealing the true picture of the initial collision.

This chapter begins by discussing the simple building blocks and technical considerations involved in the design of any modern detector. After that, details will be described for each of the major detectors arrayed around the LHC. We will concentrate on the ATLAS and CMS detectors, with fewer details given for the ALICE and LHCb detectors. The TOTEM and LHC-Forward detectors are extensions of ATLAS and CMS and will be mentioned only in passing.

Before we discuss technology and techniques, we must spend a moment talking about the kinds of particles we need to be able to detect. There are literally hundreds of kinds; however, we don't need to know about all of them to understand the most important points. The handful of particles we need to know about are the electron, the photon, the muon, the neutrino, and a class of particles called hadrons. Electrons and photons are relatively well-known particles, appearing as they do in the familiar world of human experience: electrons in electricity and photons in light. Electrons and photons do not penetrate deeply within a detector.

Muons and neutrinos are less familiar but were introduced in chapter 1. Muons are basically heavy electrons, although they interact very little with the detector and usually pass through it, leaving behind only a small fraction of their energy. Neutrinos have no electric charge, have nearly no mass, and experience only the weak force; they pass through a detector without interacting at all. In essence, they are not seen in a detector and their presence is known only by their absence.

Hadrons are a class of particles that contain quarks within them. The protons and neutrons are the most well-known hadrons, although they are relatively rare in the debris of particle collisions. The most common type of hadrons in a particle collision are called *pions,* and they can be treated in many ways as if they were light protons, having only about 15% of the mass of a proton. The manner

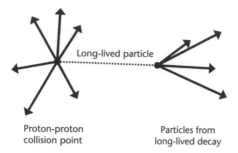

Long-lived particle

Proton-proton
collision point

Particles from
long-lived decay

Figure 4.1. Particles that live a long time will travel a considerable distance from the point of common origin or vertex. Thus when you see particles originating from a point other than the one of common origin, you are seeing evidence of a long-lived particle.

in which hadrons interact with matter is midway between electrons and muons. Hadrons penetrate more deeply than do electrons and photons but not nearly as deeply as muons and neutrinos. The different ways in which these particles interact with matter play an important role in revealing their identity.

Identifying the point of origin of a particle is often very important. This is because sometimes rare particles are made that live for a long time—several trillionths of a second. This seems very short, but highly energetic particles with this lifetime live long enough to travel millimeters or as much as a few centimeters before decaying. Since particles that live this long frequently occur in rare physical processes, you'd like to pinpoint when these kinds of particles are made. Typically, one identifies such events when the trajectory of particles in the event is reconstructed, and it becomes apparent that not all of them originated from the same point that the protons collided. When we project particles back to their point of common origin, we call this a "vertex." Vertices that differ from the collision point are of interest to physicists. Figure 4.1 shows what the signature of a long-lived particle might look like.

There are many clever techniques for discovering the identity of particles, but we only need to know a few. They have technical names but actually are pretty simple concepts. The topics we will discuss are these: magnetic bending, ionization, showering, and the rather ominous-sounding duo, transition radiation and Cerenkov radiation. We'll introduce each of these ideas in turn.

Magnetic Bending

The first of the techniques, magnetic bending, is one we've encountered already. In chapter 3, we discussed large circular accelerators, which you might recall consisted of a short acceleration region and a vast array of magnets the sole purpose of which is to guide the protons in a circular path back for another

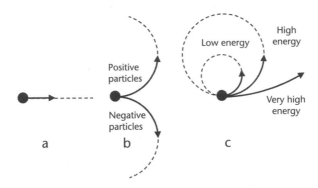

Figure 4.2. The effect of magnetic fields on different kinds of particles. *a*, A magnetic field does not deflect a neutral particle; *b*, a magnetic field deflects particles with electric charge, with particles of opposite charge being deflected in opposite directions; and *c*, particles with low energy travel the circumference of a circle with a small radius and higher energy particles follow circles of larger radii.

acceleration phase. The crucial point here is that charged particles move in a circular path when they are being influenced by a magnetic field.

This fact can be exploited to help identify and measure the charged particles coming out of a collision. A circle is a simple geometric shape. The only thing that distinguishes different circles is their size. So that having particles traveling in a circular path can be a useful technique for measuring particles, we have to be able to relate a circle's size (that is radius or circumference) to an important particle property. This turns out to be possible, and the important property is the particle's momentum. In our ordinary experience, momentum is related to the velocity of the object: the higher the velocity, the higher the momentum.

At the high energies involved in modern particle physics collisions, the correspondence between velocity and momentum doesn't hold, but it's still a valuable mental picture. However, in these ultra-violent particle collisions, momentum is more like energy. Because the term is more familiar, I will apologize to my physicist colleagues and use the term *energy* here.

Because the size of the circular path followed by a charged particle is related to the particle's energy, by measuring the size of the circle you have simultaneously determined the energy carried by the particle: the bigger the circle, the bigger the energy.

Figure 4.2 illustrates how the energy of a particle is related to the size of the circular path it follows. It also shows another interesting feature. Particles with opposite electric charge (for example an electron and an antimatter electron) curve in opposite directions. If a positive particle moves counterclockwise, a negative particle will move clockwise.

Figure 4.3 shows an example of a relatively simple particle collision. In this

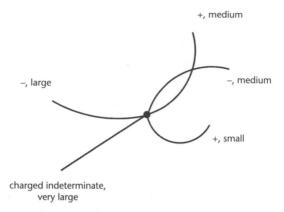

Figure 4.3. Particles originating from a common origin can be deflected by a magnetic field. In the figure, the (+/−) signs indicate the sign of the electric charge and the words denote the amount of energy they carry. Note the deflection direction and the curvature of the various tracks in relationship to their charge and energy.

example, five particles exit the collision. The particles have different energies and electric charges. From the clockwise and counterclockwise directions of the particles' motion, we see that we have two particles with positive electrical charge and two negative. The fifth particle has such a large energy that it is difficult to say if it is curving clockwise or counterclockwise. So, because of this ignorance, we are unable to say whether this particle is positive or negative. Further, this particular particle underscores an important limitation of this technique. For instance, if the particle is moving with so much energy that you can't tell whether it's moving clockwise or counterclockwise, then that essentially means that it is moving in a nearly straight line.

Moving in a straight line means that it is moving along the circumference of a huge circle. This means that it has a lot of energy. The problem is that once you get to that much energy, you can't tell if the circle is huge, super-huge or super-duper-huge. And, if you can't accurately measure the size of the circle, you can't accurately determine the particle's energy.

From this, we see that the magnetic bending technique works best when the energy of the particle isn't too big. This leads us to wonder how we can accurately measure the energy of highly energetic particles. This question is even more pressing given that we know that the LHC is the highest energy accelerator ever built. Luckily, there is such a technique called *showering* that works better as the energy of a particle increases. We will explore this technique presently, but first let's look at the second technique in our list: ionization.

Ionization

In our discussion of magnetic bending, we learned about the path of a particle and its relationship to the particle's energy. We didn't explore exactly how we see the particle. For this, we need to talk about ionization.

When a charged particle passes through a chunk of material, it bounces into the atoms in the material. Now unlike a bowling alley, in which the ball must physically hit a pin to knock it over, the charged particle is surrounded by an electric field. This electric field extends far beyond the size of the charged particle itself. This electric field can reach out and jiggle the atoms of the material through which it is passing.

This is a bit tricky to imagine, so let's think up some analogies. If you take a magnet and move it near some iron nails, sometimes the nails will be attracted to the magnet, even though the magnet doesn't actually touch them. Or one might think of a big truck moving down a street covered by newspapers and Styrofoam cups. The wind from the truck's passage will move the debris around, even though the truck never physically hits it. So too it is with the electric field surrounding a particle carrying an electric charge. As this effectively large object (the charged particle with its extended electric field) plows through material, it bounces into the material's atoms. With each bounce, the charged particle slows down just a bit, like a bowling ball rolling through a room filled with pins. Fundamentally, that's all ionization is: a charged particle moves through material, bouncing into the material's atoms and slowing down in the process.

The next thing you need to know about ionization is that the amount of energy a particle loses is proportional to the distance of matter through which the particle travels. Say the particle loses one unit of energy after traveling through a fixed length of matter (say a centimeter or an inch). Then after traveling through three times that length, it loses three units of energy. Fifteen centimeters (or inches) means 15 units of energy loss and so on. So, conversely, if you measure the distance through which the charged particle travels, you know its energy. In our example above, if a particle travels through 100 cm (or inches) of material and then stops, you know that it had 100 units of energy when it started.

So with those technical concepts out of the way, let's step back and take a look at what ionization means. For all intents and purposes, it's the same as slamming on the brakes of your car. The loss of energy as a result of ionization is effectively similar to the loss of the car's energy as a result of the friction between the tires and the road. And, just like a long skid mark means the car was moving quickly when you hit the brakes (e.g., it had a large initial energy), a charged particle penetrating deeply into matter means its initial energy was large.

So just how deeply can a particle, slowing only by ionization, penetrate into matter? Well obviously that depends on the energy of the particle and the material through which it travels. Taking a relatively low energy particle (10 GeV for the technical types) in solid iron, a particle can travel about 7.6 m (25 feet). Now given that the energy involved in an LHC collision is 14,000 GeV, particles with such a low energy will be very common. Particles with ten times as much energy will be pretty common as well. So these higher (but relatively common) energy particles would require a chunk of iron about a football field deep to stop them.

Given that modern particle physics equipment can't be that big (imagine a sphere of iron, 180 m, or 600 feet, in diameter around a collision point, costing 10 full years of the entire U.S. federal budget at 2007's price levels), there must be another solution or different technique we can use. This budget-saving technique is called showering.

Showering

While everything we've said about ionization is true, for some particles, it's not the entire story. Some particles will undergo additional types of interactions. The particles in question are electrons, photons, and hadrons (i.e., quark-containing particles). When these particles pass close to an atom, in addition to slowing down through ionization, they can actually split into two or more particles. For instance, if an electron passes close enough to a nucleus, it can kick off a photon: one particle in (electron) and two particles out (electron and photon). Similarly, when a photon comes close to the nucleus of an atom, it can disappear and be replaced by an electron and a positron: again one particle in and two out.

Now here's a nifty thing. When a particle splits into two particles, they each get (about) half of the energy. So if the distance a particle can penetrate into matter is related to its energy, this splitting has converted one particle that can go a certain distance into two particles that can go half that distance.

To appreciate showering, you need to know another interesting fact. Once the one particle becomes two particles, well then these two "daughter" particles can also hit atoms and split. In this way, one particle can turn into two, then four, eight, sixteen, and so on. Indeed, it isn't at all unusual for one particle to "shower" into ten thousand. With that increase in particle count comes a reduction in each particle's energy. Most important, showering vastly reduces the amount of material needed to fully absorb a particle and measure its energy. The basic idea of showering is shown in Figure 4.4.

The quark-carrying hadrons shower as well, although the details are a bit different. When all effects are taken into account, we find that hadron showers

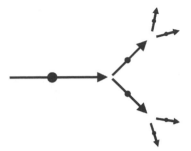

Figure 4.4. Showering. A particle will split into two particles with less energy. Subsequent splits increase the number of particles and reduce the energy of each.

last longer and penetrate more deeply. Roughly speaking, the electromagnetic electron and photon will penetrate several centimeters into a dense material like metal, while hadrons will penetrate a meter or so.

With the introduction of ionization and showering, we can get a first glimpse of how physicists start to distinguish between particle types. Let's take a simple two-component detector, with one section gaseous and in which ionization is measured and one section solid, in which showering occurs. (Don't worry about how the ionization is measured, we'll get to that later.) To show the essential points of how different particles are identified, let's consider five types: neutrinos, muons, photons, electrons, and hadrons. Neutrinos are electrically neutral and don't shower. Photons are electrically neutral and shower quickly. Electrons have an electric charge and shower quickly. Hadrons can be neutral or electrically charged and shower slowly. Finally, muons have an electric charge, but don't shower.

In Figure 4.5, we see that electrically charged particles are observed in the gaseous region, while neutrals are not. In the solid region, particles shower as their nature dictates. By looking at the patterns in both detectors, the identity of the originating particles can be determined with considerable reliability.

That Pale Blue Glow

Before we get into the specifics of detectors, we need to introduce two additional useful effects: Cerenkov radiation and transition radiation. Most people with even the smallest science interest and training know that you can't go faster than the speed of light. (Although judging from the crank letters and e-mails I receive every month, this fact is not universally accepted.) Technically, the right thing to say is that you can't go faster than the speed of light in a vacuum. However, when light travels through a material, it moves more slowly. In fact, light travels through glass or Plexiglas at about two-thirds the speed it has in a

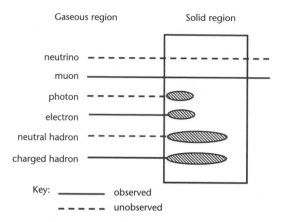

Figure 4.5. The different signatures of particles in matter make it possible to distinguish among them. By observing whether the particle ionizes in the gas or showers in the solid, one can fairly reliably identify the particle's nature. The cross-hatched ellipses indicate a shower. Observed means the detectors will be able to see this particle, while unobserved means that the particle will be invisible to that detector.

vacuum. This fact forms the basis for how lenses, prisms, and any number of optical phenomena work.

That light travels more slowly when moving through glass leaves a loophole in that whole "faster than light" thing. This is because when light hits glass, it slows down immediately (and speeds back up when it leaves the glass). However, the mechanisms that cause the slowing do not affect particles other than the photon. Thus we are left with the following situation: Suppose you had two high energy particles, one electron and one photon, traveling alongside one another in a vacuum. No matter how high the energy carried by the electron, you'd see the photon pull ahead of the electron. If the electron was very high in energy, the photon might inch ahead, but ahead it would pull.

Now send these two particles into a slab of glass. The photon would immediately slow down to about two-thirds its initial speed, while the electron's speed would be essentially unchanged. Thus in glass, the electron can be faster than the photon! When this happens, an effect occurs that is similar to a sonic boom. A sonic boom occurs when an airplane moves through air faster than sound moves through the same air. Similarly, when an electrically charged particle travels through glass faster than light travels through the same glass, it gives off light. When this light is observed, you can be sure a highly energetic charged particle has passed through the glass. This light is called *Cerenkov radiation* or *Cerenkov light*.

We can merge the ideas of Cerenkov light and showering to create a powerful particle detection tool. We will return to this idea later, but suppose you have a block of lead glass, which is used to make any high-end chandelier. Unlike ordinary glass, which is essentially sand that has been melted and cooled, lead glass not surprisingly consists substantially of lead. Recall that when an electron comes close to an atom, it gives off a photon. Further recall that a photon, passing near an atom, will split into electrons and positrons. Finally, recall that these daughter particles can also pass near atoms and that the process will repeat itself. This is the shower we discussed earlier. However, in this case, the shower grows in glass. Because the electrons in the shower are very fast (and exceed the speed of light in glass), the electrons and positrons emit Cerenkov light, which can be collected and converted into electricity for further processing. So a chunk of lead glass and a high-tech electric eye can provide a way to measure the energy of electrons and photons.

Transition Radiation

The last technology we're going to describe is transition radiation. As its name suggests, this is radiation caused by a transition. Ah, but what transition? When a charged particle travels through a medium such as glass, it is surrounded by an electric field that is determined by its own electric charge and by the surrounding medium. However, since the electric field depends in part on the medium through which it travels, the electric field will change as the particle passes from one material to another (say glass to air or plastic to liquid). In the transition from one medium to another, an x-ray photon is emitted from the charged particle. X-rays themselves are not seen, but they have enough energy to interact with the material and induce ionization. Figure 4.6 shows the basic idea.

If you carefully select the materials and the shapes of the materials, you can precisely locate where a charged particle has made the transition. Further, since transition radiation depends on a particle's velocity, this phenomenon can be used to distinguish fast particles (like the light electron) from slow particles (like the heavy, quark-carrying, hadrons). The ATLAS and ALICE detectors at the LHC use this technology, and we'll describe how in more detail when we get to the sections in which specific detectors are described.

We now know something about principles important in particle physics detectors: magnetic bending, ionization, showering, Cerenkov radiation, and transition radiation. You may also recall that we want to know as much about the particles coming out from a collision as possible, with special attention paid to their point of origin (usually the place where the collision occurred), their trajectory, electric charge, energy, and identity. It's now time to bring these

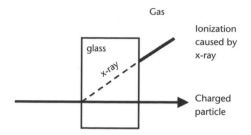

Figure 4.6. Transition radiation occurs when a charged particle travels from one kind of material to another. X-radiation is emitted at the transition and these x-rays can ionize a gas for detection.

ideas together. Obviously there are as many different possible solutions to making the desired measurement with the available technologies as there are clever scientists and engineers. Accordingly, we will restrict the discussion to those choices made as part of the design of the various LHC-based detectors.

To understand the choices one might make, we first draw a simple cartoon, with only a few kinds of particles. Like the earlier showering discussion, we will include an electron, a photon, a positron (an antimatter electron with the opposite electric charge of an electron), a muon, a neutrino, and both electrically charged and neutral hadrons.

Figure 4.7a shows these particles, when you know everything about them. The identity and electric charge of each particle is given, as is how much energy each one carries. This is what physicists call the *truth level*. But, of course, our detector doesn't provide us with this perfect knowledge. In the following paragraphs, we are going to apply some specifics about our newly learned detector knowledge and get an idea of what physicists actually see.

Recall that we want to know the electric charge and energy of the particles. One of the techniques we discussed was magnetic bending. Magnetic bending makes particles with electric charge move in a circular path. The size of the circular path is related to the energy the particle carries. Further, particles with positive electric charge curve in the opposite direction of negative particles.

So let's apply a magnetic field to the particles of Figure 4.7a and see what effect it has. Figure 4.7b shows how the example particles react to a magnetic field. The low-energy electron and positron are bent a lot. The positively charged hadron with medium energy is bent a middling amount, while the trajectory of the high energy muon is bent only minimally. The other particles, being electrically neutral, are not affected by the magnetic field.

We've made our first steps toward understanding a particle scattering collision, but there's just one obvious problem. We've not actually detected the passage of the particles. To do that, we need to dig into our bag of tricks. To view the particles' paths, we need to use ionization. Recall that ionization occurs when an electrically charged particle crosses through a material and interacts with the material's atoms. The effect of the particle's passage is then detected via various methods.

Typically in an ionization detector, you'd like to minimize the total amount of material. If you have too much material, other effects we've not discussed (and won't) come into play and things get complicated rather quickly. So to minimize the amount of material through which the particles must pass, ionization detectors consist of many layers of material, separated by a low density material, such as a gas or vacuum.

In our simple example, we surround the collision with a series of concentric circles of material. Figure 4.7c shows what happens when the detectors are added. Electrically charged particles ionize the material and their passage is recorded, while the neutral particles slip through unscathed. In the figure, the passage of each charged particle through the ionization detector is recorded by a little dot.

So far, we've been able to detect electrically charged particles but not the neutral ones. To detect them, we need to add the next trick: showering. Recall that showers provide a way for particles to dump all of their energy rather quickly in a dense material. Electrons, positrons, and photons have short showers, while hadronic particles have longer showers.

Shower detectors are usually comprised of thick slabs of dense material, usually metal. Thick is important, because if the detector is too shallow, the shower might leak out the other side and that means energy would be undetected.

In Figure 4.7d, we see the effect of adding showering detectors. The depth of a shower depends mostly upon the identity of the particle that is causing it. We note that the neutrino has yet to leave a trace in any detector. Further, the muon doesn't shower and passes through the dense material, leaving only ionization energy. Since the magnetic field isn't found in the metal, the muon travels in a straight line there. To make sure the particle is a muon, modern detectors typically have a few additional ionization-based detectors outside the showering detector. There may or may not be a magnetic field where the outer ionization detectors are situated. In our simple example, let's put a magnetic field there.

Figure 4.7e shows this final detector configuration, with knowledge removed of the charge, energy, and identity of the particles that left the signals. Contrast this with Figure 4.7a, where the truth information was revealed. Modern particle experimental physicists train to turn what they can detect (Figure 4.7e) into what was there in the beginning (Figure 4.7a).

So far, we've been describing detectors in general terms. Now it is time to get more specific. Our essential questions must be the following: How do we mea-

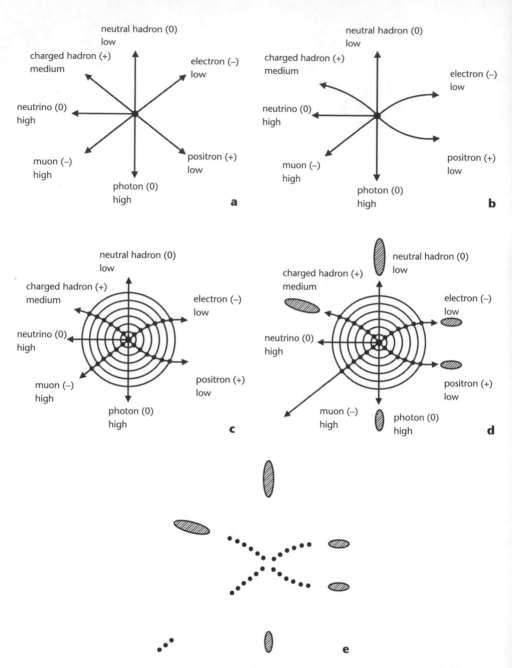

Figure 4.7. *a,* After a particle collision, particles exit the vertex or point of common origin. Denoted here are the identity, electric charge, and energy of the various particles. *b* is the same as *a* but with the effect of a magnetic field imposed. *c* is the same as *b* but with a typical ionization detector superimposed. The filled-in dots show where the charged particles traverse the detector and leave a signature. Note that the neutral particles are not observed. *d* is the same as *c* but with the showers added. The cross-hatched ellipses show a shower, with the size of the ellipse showing how deep the particles penetrate. Hadron-initiated showers penetrate more deeply than electromagnetically initiated ones. *e* is the same as *d* but with all extraneous information removed. Note in three additional ionization detector hits (*bottom left*), which are from the muon ionization detection system that is outside the shower-measuring device. Compare panel *e* (which is what is available to the experimenter) with *a* (which is what we are attempting to observe).

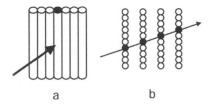

a   b

Figure 4.8. A typical ionization detector. Individual detector elements are traversed by a charged particle and signal the particle's passage. *a*, face-on view; *b*, an edge view of multiple planes.

sure the position of a particle? What are the essential components of a showering detector? Let's start with the first question.

Ionization Detectors

The basis for most position-measuring detectors is ionization. Electrically charged particles cross some material and disturb the electrons in the atoms of the material, and the disturbance is somehow detected. The trick is then to find out where the particle enters the material. While there are a number of choices one can make to accomplish this, by far the most common is to simply make many small ionization detectors, isolate them from one another, and just use your information about which ones were hit to tell you where the particle passed.

As an example, suppose you made a detector somehow composed of ordinary soda straws. Inside each straw is some unspecified material that can be ionized and in which the ionization can be detected. If you took these straws and laid them side by side, they would form the plane seen in Figure 4.8.

Each straw tells us when it is hit. We don't necessarily know where in the straw the particle passes, but at least we know which one and that tells us something about the particle's position. If we have many planes of these straws, we can then measure the particle's path. In Figure 4.8b, we can view many of these planes edge-on and see how the pattern of hit straws gives us the information we need.

Figure 4.8 illustrates an important point. Although the whole "straw" technique can give the necessary position information, a big limitation is the size of the individual straws. Bigger straws measure the position less precisely, while smaller straws give more precise measurements. So it is clear that you should make your straws as small as technologically possible, right? The answer is "Yes, but . . ." This important "but" reminds us of the real-world consideration of cost. Each straw needs its own electronic circuit to read it out, and each circuit adds a cost to the budget. So more straws mean more expense. The reality of fixed budgets imposes a very real limitation on how small one can make the individual straws. In the end, scientists compromise. When precise measurements are critical, they make small detectors and accept the large cost, but when circumstances allow, they use bigger straws and spend their money elsewhere.

In our example, we used a hypothetical detector made of straws. Although there are detectors literally made of straws, that technique is relatively rare. More commonly, scientists use two techniques: wire technology and silicon technology.

In wire technology, the straws are replaced by wires. In fact, a plane of wires looks a lot like a harp. The wires are placed inside a container filled with a carefully chosen gas, and the wire nearest the point where the particle crosses the plane is the one that reports the particle's passage. The space between adjacent wires varies depending on detector design, but a centimeter or two (a quarter or half an inch) is reasonable. With sophisticated electronics, detectors using this technology can be made to measure a particle's position with a precision of about a few tenths of a millimeter, or a hundredth of an inch.

To measure more precisely, one must turn to silicon technology. In recent decades, engineers have made enormous strides in minimizing the size of electronic chips for use in computers. This technology can be turned to making particle detectors. In silicon detectors, the "straws" are little strips of silicon, a few centimeters (a couple of inches) long and so narrow that you could fit 20 of them in the space of a millimeter. Recently, it has become economically and technically feasible to make little square "dots" of silicon, 0.05 millimeters on a side. (Although instrumenting these tiny detectors is extraordinarily expensive.)

As I mentioned above, there is a type of detector that uses something that really does look like straws. In this case, charged particles traverse literal straws, filled with material that experiences ionization. As charged particles cross the straws, transition radiation is emitted in the form of x-rays. These x-rays then ionize the material inside the straws and so the particles are detected. The ATLAS detector uses this technology.

Showering Detector Techniques

With our discussion of ionization detectors complete, we turn our attention to showering detectors. Showering detectors all have one thing in common. They all consist of dense material. Two techniques are the most common, and the first is called a *sampling calorimeter*. Its name comes from the fact that it measures a representative sample (sampling) of a particle's energy (calorimeter). The basic structure of such a detector consists of a series of metal plates, separated by a material in which it is easy to measure ionization energy. This material can be a gas, solid, or liquid, with a density that is lower than that of metal. A typical detector of this form might consist of a plate of steel a few centimeters (an inch)

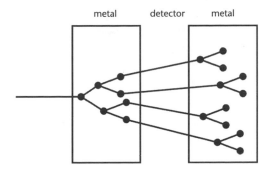

Figure 4.9. In a real sampling calorimeter, the bulk of the showering occurs in high-density metal, while the actual detection occurs in the low-density material between the metal plates.

thick, followed by a centimeter or so (half an inch) of the material that actually detects. This pattern might repeat 50 times. The details (thickness and materials used) of any particular detector will vary.

Figure 4.9 illustrates the important features of a sampling calorimeter. Particles interact with atoms in the high density metal and shower. These shower particles then travel through the active detector material, where the ionization is recorded. The showering begins again in the next metal layer. As this pattern repeats itself, the energy of the individual shower particles drop. Eventually, the energy of the particles drop enough that showering stops. These particles are then simply absorbed and the shower ends. The entire process takes a few billionths of a second.

The second kind of showering detector is called a *homogeneous calorimeter*. Unlike a sampling calorimeter, a homogeneous calorimeter doesn't have any structure. The whole detector is the same. To have a detector that can be homogenous, contain metal, and be able to be read out is quite a trick. This is usually accomplished by using some kind of metal-containing glass. The most common form of this kind of glass is lead crystal. That's right, the same crystal that makes up the chandeliers in a chic hotel's grand ballroom or in that decanter your mom got for her wedding is an ideal material to be used in a particle detector.

A detector made of lead glass works slightly differently from a sampling calorimeter. A high energy particle enters the glass, traveling faster than the speed of light through the same material. This particle emits Cerenkov light. The particle encounters a lead atom and showers. The daughter particles after the shower are also traveling faster than the speed of light in the glass, so they too emit Cerenkov light. These daughters also encounter lead atoms and the shower grows. The daughter (and granddaughter, and on and on) particles all emit Cerenkov light. Because the detector is made of glass, the Cerenkov light travels to the

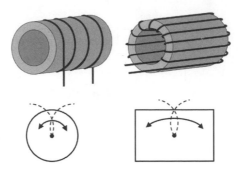

Figure 4.10. Types of magnets: solenoid (*top left*) and toroid (*top right*). A solenoidal magnetic field will bend charged particles either clockwise or counterclockwise when the solenoid is seen face-on (*bottom left*). A toroidal magnetic field will bend the trajectory of charged particles in a plane that includes the side view of the toroid (*bottom right*). Courtesy Barry Panas.

end of the detector and is collected and converted into electricity. These kinds of metal-containing glass particle detectors are predominantly used to detect electrons, antimatter electrons, and photons.

The "Oids:" Solenoids and Toroids

Because two of the LHC detectors have unfamiliar words in their name, we must make a brief detour into how specific magnetic fields can be made. All magnets in modern particle physics are made by wrapping coils of wire in various shapes. Electric current is passed through the coil, and that current is what generates the magnetic field. These shapes form different kinds of magnetic fields, which in turn bend the particles in different directions. The two most common types of wire-wrapping patterns are the solenoid pattern, in which the wire is wrapped around the outside of a cylinder in the shape of a spring or a "slinky." A toroid pattern is formed by wrapping the wire in the shape of a bagel. A solenoidal magnetic field bends particles in the plane perpendicular to the beam, while a toroidal magnetic field bends particles toward or away from the beam. Figure 4.10 illustrates these slightly confusing words.

Specific Detectors at the LHC

With our discussion of detector techniques complete, we are now able to outline the specific detectors arrayed around the LHC. When the accelerator turns on, there will be two general-purpose detectors designed to study the highest energy proton-proton collisions (CMS and ATLAS). Two additional detectors are designed to study much lower energy phenomena (LHC Forward and Totem). These detectors can be considered to be "add-ons" to the two main detectors. In addition, there will be two other detectors that have been carefully designed to answer much more focused questions. The LHCb detector is optimized to study collisions in which bottom quarks are produced, while ALICE is designed to study the collisions of heavy ions, for instance when lead nuclei are collided together at nearly the speed of light. So let's introduce each detector in turn.

The rest of this chapter discusses specific detectors stationed at the LHC. Because we are being specific, there are a lot of details given, specifically sizes and number of pieces that go into the individual detectors. If you're not a detail kind of person, you can skip the rest of this chapter. The principles discussed thus far are featured in various ways in all of the LHC detectors and so you can have a pretty good, although general, idea how all of these detectors work with what you've learned to this point. For those of you who are detail-oriented, let's plunge ahead. We'll meet up again with the "big picture" people at the start of chapter 5.

Compact Muon Solenoid (CMS)

The Compact Muon Solenoid, or CMS, detector gets its name because it is relatively small (Compact), is optimized to study muons (Muon), and has a solenoidal magnetic field at its heart (Solenoid). Of course, compact is relative. It's three stories tall. When we contrast the ATLAS and CMS detectors, the basis of the "compact" term will become more apparent.

Like most modern particle detectors found attached to particle accelerators, CMS (and ATLAS and ALICE) have a layered "onion" structure, with different types of detectors nestled within one another. For simple reasons of engineering, the layers in these detectors are roughly cylindrical in shape, looking like nothing more than a large soup can. The various cylinders representing each type of detector technology are nested together like a series of Russian matrioshka dolls.

Figure 4.11 shows us the CMS detector. It is 19.8 m (65 feet) long and 14.6 m (48 feet) in diameter and weighs 12,500 tons. It consists of six distinct layers in the "central" or "barrel" region (e.g., the sides of the soup cans) and five distinct layers in the "end cap" regions (e.g., the top and bottom of the can). In the barrel region, these layers consist of the following: two different types of silicon detectors; a calorimeter to measure the energy of electrons, positrons, and photons; a calorimeter for measuring the energy of the quark-containing "hadronic" particles; a magnet; and finally a system for observing muons. The end cap detectors have the same structure except without the magnet. So let's learn something about these various detectors. The language can be a bit confusing. The equipment that makes up an entire particle physics experiment is called a "detector." Each experiment consists of many different subsystems, each tasked with a particular purpose. While each subsystem is properly called

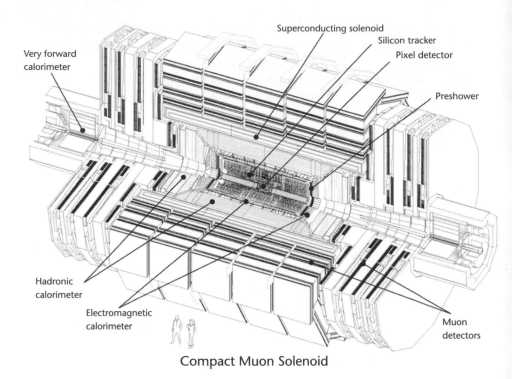

Superconducting solenoid

Silicon tracker

Pixel detector

Very forward
calorimeter

Preshower

Hadronic
calorimeter

Electromagnetic
calorimeter

Muon
detectors

Compact Muon Solenoid

Figure 4.11. A view of the CMS detector, with the various pieces identified. Note the size of the people (drawn to scale) at the bottom. Courtesy CERN and the CMS collaboration.

a "subdetector," the word "detector" is often used for the overall system, as well as the respective subsystems.

The silicon tracking detectors of CMS are simply staggering in their technical parameters and are currently without peer. The silicon tracking detectors sit in a volume about 5.8 m (19 feet) long and just under 2 m (a little over 6 feet) in diameter. This volume is not entirely filled with silicon but rather with layers of silicon and air. Each layer of silicon is a fraction of a millimeter thick and is mounted on a cylinder made of a carbon fiber composite. Carbon fiber is used to make modern ultralight airplanes because of its strength. Adjacent cylinders are separated by few centimeters (an inch or two).

The silicon system is broken up into two different subsystems, one consisting of tiny silicon detectors, while the other system consists of super tiny ones. The inner system is called the *pixel detector,* so-named because it contains super tiny pixels of silicon. To get an idea of what a pixel means, imagine an old-style television. If you were to look closely, you would see that the screen contained little dots, or pixels.

The pixels in the CMS silicon system are much smaller than the TV ones. If one were to look at the pixel detector under a microscope, you'd see that each square millimeter contained about 60 distinct detectors. With such tiny granularity, the CMS pixel detector consists of about 66 million distinct pixels. This incredible number of detectors is spread over a mere three cylinders, about a meter (3 feet) long and ranging in radius from a little under 5 cm (2 inches) to 10 cm (4 inches).

The second CMS silicon detector is much larger. It is 5.8 m (19 feet) long and consists of 10 cylinders ranging in radius from about 20 cm to 1 m (8 to 40 inches). Being so much larger, you'd expect that the number of silicon detectors involved would be much larger, but this apparatus consists of "only" 10 million individual detectors. (I don't know about you, but to use the words "only" and "million" in the same sentence seems weird to me.) The reason that this system contains so few individual detectors is that each one is much longer. They are called *silicon microstrip detectors,* so-named because each is a thin strip of silicon about 0.15 mm wide and about 10 cm (a couple of inches) long. Had we lived in a world without resource limitations, we'd have liked to have made just one kind of detector, consisting only of pixels, but cost prohibited that option. So, as we learned in our general discussion of silicon detectors, you make a finely grained detector when you must and make one with larger individual elements when you can.

The CMS silicon detectors cover an enormous area. If you took all of the silicon comprising the CMS detectors and laid it edge to edge, it would entirely cover the floor of an average two-story American house, with just about enough space left over so you could stand and enjoy your handiwork.

For the same reasons that your computer needs a fan, the CMS detector needs to be cold to run well. When silicon is warm, the silicon will generate within itself an unacceptable amount of electric current and stop working. Further, we must use electricity to make the silicon detectors work. Like most simple household appliances, the silicon heats up when powered. When the silicon is working properly it will be operating at about −12°C (10°F).

The next detector one encounters as one moves outward from the center is an unorthodox choice. Ordinarily, the next layer would be the coils through which electric current flows to make the magnetic field. However, in CMS the next layer is the calorimeter used to measure the energy of electrons and photons. Since photons and electrons are electromagnetic particles and the device used to measure energy is called a calorimeter, this device is called the electromagnetic calorimeter, or ECAL.

The ECAL is an example of a homogenous calorimeter. Rather than the lead glass that was discussed in the overview section, the CMS ECAL is made

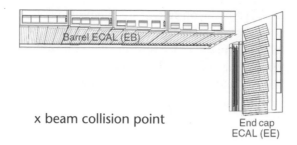

x beam collision point

End cap
ECAL (EE)

Figure 4.12. A side view of the electromagnetic calorimeter (ECAL). Only a quarter of the detector is shown. The rectangles denote individual lead-tungstate blocks. Courtesy CERN and the CMS collaboration.

of blocks of lead tungstate ($PbWO_4$ for the chemically minded). Lead tungstate is amazing stuff. While a casual inspection of one of the blocks used in CMS would lead you to believe that it was ordinary, if rather clear, glass, each block is actually 98% metal by weight. Each block is about 2.5 cm (1 inch) in height and width and about 25 cm (9 to 10 inches) long.

Figure 4.12 shows how these blocks are arrayed around the collision point. The ECAL basic shape is a cylinder, with blocks around the barrel and on the end caps. The barrel requires 61,200 blocks and the two end walls 14,648 for a grand total of 75,848 blocks. Taken together, the lead tungstate blocks in CMS weigh about 90 tons.

The next layer in the CMS detector is the hadronic calorimeter, or HCAL. Recall that hadrons are particles containing quarks, of which the proton and neutron are the most familiar. In particle physics collisions, the most common hadrons are pions, which are essentially light protons.

The HCAL is a sampling calorimeter. Like the ECAL, the HCAL is cylindrically shaped with a barrel and ends. The metal most used in the HCAL is brass, although steel is used in a couple of places. Recall that a sampling calorimeter requires layers of metal interspersed with layers of material in which the ionization energy is measured. In the HCAL, this low-density material converts the ionization to light, which is converted in turn to electrical signals. Mostly, the light-producing material is a type of plastic—very similar in appearance to Plexiglas. The layers of metal and plastic consist of plates of brass, 5 to 8 cm (2 or 3 inches) thick, followed by about 0.3 cm (0.125 inches) of plastic.

Recall that the ECAL was made of many blocks of lead tungstate. By knowing which block was hit, you could determine the position where an electron or photon hit the ECAL. The HCAL is conceptually pretty similar, with stacks of metal and plastic effectively making blocks. In CMS, there are 2,592 blocks in the barrel of the HCAL and 2,592 blocks on the ends. In addition, there is an-

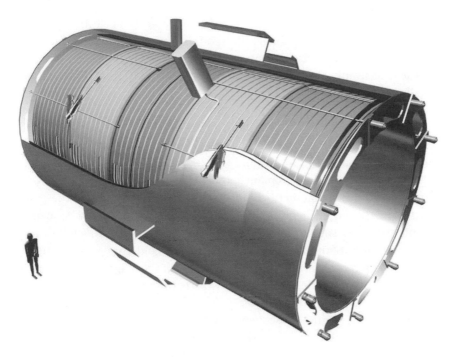

Figure 4.13. The CMS solenoid magnet, the largest one ever made. Courtesy CERN and the CMS collaboration.

other calorimeter very near the beam. This calorimeter is made of steel to make the showers and quartz to make the measurement.

Because CMS made the unusual choice to put all the calorimeters inside the magnet, the HCAL isn't quite as thick as it should be. This is because to make it thicker, the magnet surrounding it would need to be bigger. Since cost considerations made that choice impossible, a few more layers of calorimeter were added onto the outside of the magnet to catch the "tails" of the hadronic showers. The tail of a hadronic shower consists of the few rare particles that penetrate more deeply than usual. This "add-on" calorimeter is aptly called the "tail catcher."

Between the HCAL and the tail catcher is the CMS magnet, shown in Figure 4.13. The CMS magnet consists of a cylinder with an inner radius of 2.9 m (9.5 feet) and an outer radius of 3.2 m (10.5 feet). The cylinder is about 12 m (40 feet) long and is wrapped 2,168 times with wire. This wire carries the electric current needed to make the magnetic field. The magnetic field in CMS is very strong, about 80,000 times that of the Earth.

To make such a huge magnetic field, about 19,500 amperes of current must pass through this wire. In contrast, most houses need less than 100 amperes.

Thus the CMS magnet alone uses over 200 houses' worth of electricity or about the same as a small suburban neighborhood. To have that much electric current and magnetic field requires an enormous amount of energy (2.7 billion joules for the technically minded), or about enough energy to melt 18 tons of gold. Finally, to keep the wire from vaporizing under the onslaught of that much current, the wire of the magnet needs to be made superconducting. Superconducting, as we recall, means electric current flows without resistance (and thus without heating up the wires). To make the wire superconducting requires it to be cooled to about –269°C (–455°F).

All these technical requirements posed a serious challenge for the CMS engineers. Let's think a moment about some of the implications of these numbers. With the current required, special accommodations must be made for the power, with a special substation to power the CMS site. In addition, even though the wires of the magnet must be extremely cold, the outside of the magnet must be at room temperature. This means that the magnet must essentially be a large thermos bottle, different only in size from the one that keeps a construction worker's coffee hot.

Another engineering consequence of the design of the CMS magnet has to do with an inherently self-destructive aspect of designing a large electromagnet. Current makes the magnetic field. However current in a magnetic field feels a force. That's how electric motors work. So here we have wires that carry current that make a magnetic field. They in turn feel a force and thus want to move. The force is about 2 to 3 tons for every 0.3-m- (foot-) long segment of the wires that make up the coils. Thus to make the superprecise magnetic field, each segment of wire must withstand the force equivalent to the weight of a mid-sized car. Now recall that the coils of wire comprise 40 km (25 miles) of wire, and you get an idea of the kinds of distorting forces present in the CMS magnet.

Outside the magnet is a series of muon detectors. Because all particles except muons are stopped by the calorimeters, the environment in the muon systems is relatively benign. Unlike the detectors closer to the beam, into which thousands of particles plow every 25 billionths of a second, the muon detectors see only a few.

But for all that, the muon system is still challenging. By far, the muon system is the largest of all of the subdetectors. Its barrel region ranges in radius from about 3.5 to 7 m (12 to 24 feet) and is about 6 m (21 feet) long. The end caps range from 5.5 to 10 m (18 to 34 feet) long and in radius from about 1.4 to 6.7 m (4.5 to 22 feet).

The muon system consists of four thick slabs of iron, interspersed by four layers of position-measuring ionization detectors. Each of these four layers con-

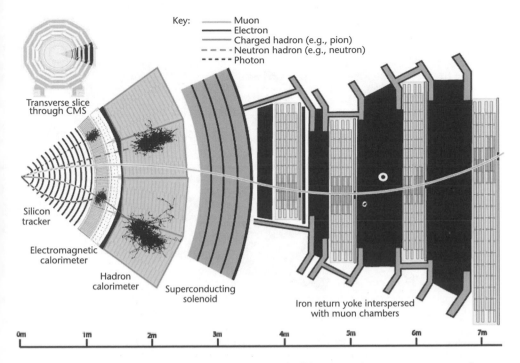

Figure 4.14. An edge-on view of the CMS detector with all important detector components and the response of the detector to various types of particles. The path of the muon curves clockwise to the left of the solenoid and counterclockwise to the right after the solenoid because the direction of the magnetic field has changed. The bottom scale indicates the size in meters. The squiggly lines show realistic showers in the electromagnetic and hadronic calorimeters. Each line shows an individual particle in the shower. Courtesy CERN and the CMS collaboration.

sists of many smaller layers. When all of the individual detectors in the muon systems are counted, they number about 830,000. Figure 4.14 shows a slice of the CMS system, drawn to scale.

ATLAS

The second big detector at the LHC we will discuss is ATLAS (for A Toroidal LHC ApparatuS). The intent of this detector is to study the same kind of collisions as the CMS, with different design choices. Given that nobody knows what kinds of new physical phenomena will be found at the LHC, it seemed prudent to have two competing detectors using different technology.

While the two detectors have appreciable differences, they also have some similarities. This isn't so surprising, since both detectors are designed to do the same thing: sit at a spot where two beams of protons intersect at their heart and

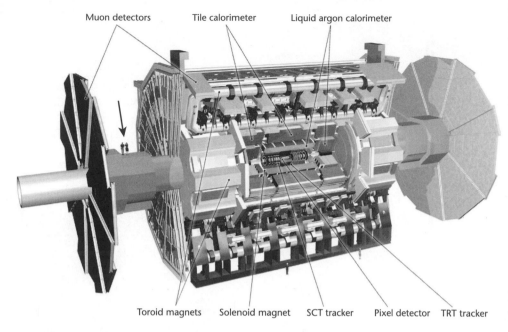

Muon detectors   Tile calorimeter   Liquid argon calorimeter

Toroid magnets   Solenoid magnet   SCT tracker   Pixel detector   TRT tracker

Figure 4.15. The ATLAS detector with components identified. Note two tiny humans on the left on what looks like an axle between the two muon detectors and two more at the base of the detector (drawn to scale). Courtesy CERN and the ATLAS collaboration.

sift through the 20 collisions that occur every 25 billionths of a second and look for something never before seen. So broadly, both detectors are large cylinders, with barrels and ends. At the heart of both detectors are silicon detectors, sitting in a magnetic field. This silicon is surrounded by the energy measuring calorimeters, followed by muon detectors.

For all their similarities, the two detectors are quite different in detail. The first difference is the physical size. ATLAS is much larger than CMS. CMS is 19.5 m (65 feet) long and 14.5 m (48 feet) wide, ATLAS is 45 m (150 feet) long and 22.5 m (75 feet) wide. So ATLAS' volume is about six times larger than that of CMS. However, even though the detector is much larger than CMS, it is also much lighter, with a weight of about 7,000 tons (compared with CMS's 12,500 tons).

ATLAS is shown in Figure 4.15. Its much larger size stems from the designers' choice to focus on muon measurements. ATLAS's muon detectors can operate alone, while CMS requires both the muon detectors and the silicon tracker to measure the properties of muons created in its particle collisions. And, as they say, time will tell which choice was best.

The center of the ATLAS detector also consists of silicon pixels. These pixels are about 0.05×0.4 millimeters in size. The pixel detector consists of three layers, spread out in a cylinder ranging from about 5 to 25 cm (2 to 10 inches) in radius and a little more than a meter (about 4 feet) long. The ATLAS pixel detector consists of 80 million pixels, somewhat more than CMS's 66 million.

Outside the volume filled with the pixel detector, ATLAS's engineers have chosen to put another silicon-based detector. Like CMS, the size of the individual silicon detectors is much larger in this region. These silicon strips are 0.08 mm wide, but about 13 cm (5 inches) long. This detector consists of about 6 million individual silicon detectors and has a cylindrical volume ranging from a radius of 0.3 to 0.6 m (1 to 2 feet) and about 5.5 m (18 feet) long.

Up to this point, the ATLAS and CMS detectors are broadly similar. However, while the CMS detector contains another silicon-based detector with a 1.3-m (40-inch) radius, the ATLAS group chose to use a different technology to fill in this volume.

The next technology encountered as we travel outward from the center of the ATLAS detector is the called the *transition radiation detector*. The transition radiation detector has a radius of 0.60 to 1m (24 to 41 inches) and is about 5.5 m (18 feet) long. Its basic construction consists essentially of long straws, four millimeters wide and about 0.7 m (28 inches) long. Eight of these straws placed end-to-end cover the entire length of the volume, and filling the entire volume requires 350,000 straws.

These straws are filled by a gas mixture that is mostly xenon. As charged particles cross the straws, they ionize the gas and are detected. However, for very fast particles (usually electrons), x-ray transition radiation is also emitted. This x-radiation also ionizes the xenon gas, leaving an even bigger signal. By seeing which straws are hit, one can follow the trajectory of charged particles through the volume. By seeing which ones have higher or lower signals, one can determine which trajectories are caused by electrons.

While the designers of the CMS detector made the unusual choice to follow the tracker with the lead tungstate ECAL, the ATLAS group made a more traditional choice. The next layer in the ATLAS detector is the central solenoidal magnet, which is similar to that of CMS. The magnet has a radius of a little more than a meter (about 4 feet) and is about 10 cm (4 inches) thick and 5 m (17 feet) long. The wires used to carry the current to energize the electromagnet wrap 1,173 times around the outside of the cylinder and carry a little under 8,000 amperes of current. The net result is a magnetic field of about 40,000 times that of the Earth, or half the magnetic field at the heart of the CMS detector.

Following the ATLAS central magnet are the two calorimeters, the electromagnetic and hadronic. Both calorimeters consist of the usual barrel and end

Table 4.1 A comparison of the various major detectors at the LHC

Characteristic	ATLAS	CMS	ALICE	LHCb
Weight (tons)	7,000	12,500	10,000	4,300
Height (ft)	70	48	51	32
Length (ft)	147	77	83	64
Price (million $)	460	460	125	63
Magnetic strength (relative to Earth)	40,000	80,000	10,000	40,000
Energy (TeV)	14	14	1,150	14
Brightness (relative to ALICE)	10,000,000	10,000,000	1	100,000
Number of collisions/second	800,000,000	800,000,000	8,000	40,000,000

Note: TeV = one trillion electron volts.

cap geometry. The electromagnetic calorimeter is of the sampling style and uses lead to create the showers and argon, chilled to a liquid form, to measure the shower energy. The ATLAS electromagnetic calorimeter consists of about 175,000 individual detectors, more than double that of CMS.

Like in the CMS, the ATLAS calorimeters used to measure hadrons are also sampling calorimeters. In the barrels, layers of iron and ionization-detecting plastic are interleaved, while in the end caps, the structure is copper interleaved with liquid argon. About 19,600 individual detectors make up the ATLAS hadronic calorimeters.

It is when we turn our attention to the final layers of the ATLAS detector, the muon detectors, that we see the greatest contrast with the CMS detector. To begin with, we finally encounter the large toroid magnets that are featured so prominently in ATLAS's name. The outer ATLAS magnets are enormous. In the central barrel region, the magnets are 24 m (80 feet) long and have a radial distance from 4.6 to 9.8 m (15 to 32 feet). The end cap toroids are 4.8 m (16 feet) long and fill the radial volume from 0.75 to 5.5 m (2.5 to 18 feet). In both sets of magnets, the current is 20,500 amperes. These are big magnets.

The ATLAS muon detection system is similarly impressively large. Various detector technologies record the ionization caused by the muons' passage. About 1.1 million individual detectors comprise the ATLAS muon system. Figure 4.16 shows a slice of the ATLAS detector.

The two large general purpose detectors at the LHC are amazing, both as feats of engineering and technology. Both detectors are designed to search for new physical phenomena hidden in the deluge of more pedestrian collisions between protons. Table 4.1 summarizes the main points of each detector, each containing nearly a hundred million distinct detector elements. Only time will tell if one group has made better design choices than the other. If history is any

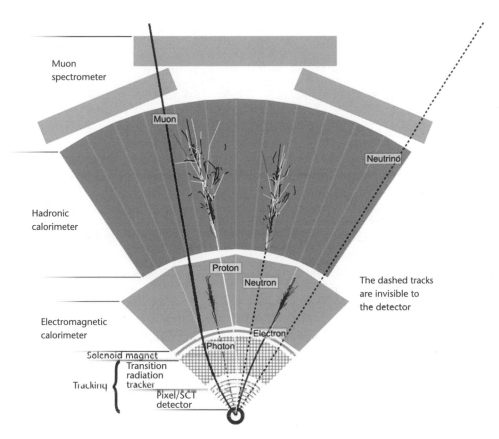

Figure 4.16. An edge-on slice of the ATLAS detector with the important components shown and how they respond to various particle types. Courtesy CERN and the ATLAS collaboration.

guide, both detectors will have an edge over the other in some particular measurements and yet both will make competitive (and superb!) measurements.

Triggers

One thing we've not discussed is the rate at which the two large detectors can collect data. Recall that the proton beams are set up to allow collisions to occur in the center of both detectors every 25 billionth of a second. That means collisions occur 40 million times each second. Further, the beams are so intense that we expect 20 or so collisions every time they pass through one another. That means that there are about 800 million collisions per second in each of the ATLAS and CMS detectors. As my youngest son told me, "Daddy, that's a lot!"

It turns out that each experiment can record and process about 100 to 200 events per second, a far cry from the 40 million. Roughly speaking, each detector can record only one collision out of every hundred thousand.

Another feature one must consider when attempting to record collisions that are of interest to particle physicists, is that they are extremely rare. Most collisions at the LHC will be relatively gentle impacts between two protons passing by one another, like two strangers brushing shoulders as they pass one another on a street in New York City. However, like the beginning of many a light romantic comedy, in which two hurrying people run head-on into one another, occasionally two protons collide violently, and some interesting physical process is revealed.

The problem is that at the LHC the gentle collisions are about 100 trillion times more likely than the ones we are interested in. Combined with the fact that the LHC experiments can record only one event out of a hundred thousand, this means that one has to be very careful in selecting just what collisions to record. The process whereby one selects events is called a *trigger* and is crucial to running a successful experiment. If you choose the wrong collisions to record and you don't have the right data to analyze, you might as well pack it up and go home.

Triggers in a particle physics context are very complex and fluid, so it is impossible to describe them in detail here. The two experiments have made different choices in their initial planning phases, and it is a certainty that by the time you read this, the triggers will be different in detail from what the groups are thinking as I write this. However, there are some essentials that will remain.

The essence of triggering is having a multiple level scheme. Experiments have two to four levels. The basic idea is that data flow into electronics (either custom-built or off the shelf computer components). These electronics are programmed to evaluate the data, decide if they are "interesting," and pass along for recording the subset of data that seems like it might be worth keeping. Each level makes ever-more-sophisticated decisions.

As an example, we can think what a two-level trigger might be. Level 1 might look to see if detectors in the muon-detection system were hit by the passage of a charged particle. If you're interested in physical phenomena that produce a muon, well then you can immediately exclude recording collisions in which the muon detectors are silent. The level 1 trigger will make this decision and pass on the events that fit the criteria to level 2.

If it turns out that the muon systems have indicated the passage of a charged particle, this doesn't necessarily mean that you want to record the event. Recall that the fraction of events that you can record are very small. So the level 2 trigger will look at the subset of events that passed level 1 and stare at them a little harder. Since the number of events entering level 2 is now relatively low, it can spend more time and determine the muon direction, energy, and other characteristics. Then level 2 will decide if the muon passes your criteria and will tell the electronics to discard the event or record it to the computer or other device.

The actual trigger system is much more complicated and looks at different facets of the collisions. But the important points are the same as discussed above. Roughly speaking, the 40 million collisions per second are presented to level 1 for consideration, and about 10,000 events per second are selected as being potentially interesting. Level 2 looks more closely at these 10,000 events and chooses 100 to 200 to record. Later, scientists study these collisions in detail, hoping to see something interesting.

Thus we see that the trigger system is a crucial piece of the design of an experiment. Just like a poor choice in the energy or particle types in your accelerator, or a poor technology choice in your detector can make your experiment a failure, so too can a poor trigger choice. The number of right choices one must make to simply record the data is rather daunting.

The Special Purpose Guys

The ATLAS and CMS detectors are the large multipurpose detectors that were the primary reason the LHC was built. However, there are other detectors at the LHC, two of which we'll mention only in passing, and two of which we'll discuss in a little more detail.

While these first two detectors are attempting to study the rare collisions that may signal new physical phenomena, these particular collisions do not make up the bulk of collisions that occur. Recall that the "interesting and rare" information is about one part per 100 trillion of the collisions. Some scientists are more interested in the common. After all, nobody has measured these common processes at these energies before.

One of the most common things that can happen when protons collide is they act like two billiard balls, just bumping into each other. Two protons enter the collision, and two exit. Because these collisions are relatively gentle, the protons are not scattered at large angles, and so don't hit the ATLAS and CMS detectors.

Thus two "add-ons" were built that piggyback on ATLAS and CMS. These are small ionization detectors, located near, but outside of, the two big detectors. These detectors record the passage of protons gently bumped out of the beam pipe. The TOTEM detector is associated with CMS, while the equivalent for ATLAS doesn't yet have a snazzy name. In addition, associated to the ATLAS detector is the LHC forward, or LHCf, detector. This detector is located about 165 m (550 feet) from ATLAS and is designed to look at neutral particles generated very near the beam. These data will help understand common occurrences when protons slam into one another and the cosmic ray collisions discussed in chapter 2. These detectors are very small (a few cubic meters) and are situated a few to hundreds of meters (tens to hundreds of feet) from the big detectors and oriented on the beamline.

Two other major detectors at the LHC remain. While the ATLAS and CMS detectors are designed to be general purpose and versatile, general purpose usually means compromise. Being able to do everything usually means that you don't do any particular thing as well as you would if you focused on it exclusively.

LHCb

The ATLAS and CMS detectors are designed to run in the punishing collision environment of having 20 or so proton-proton pairs collide at the same time. These detectors must sift through the debris, looking for some rare physical phenomena. Further, this process repeats itself 40 million times a second. The reason one would design an experiment to run under these conditions is that new physical phenomena are very rare, say one interesting collision for every one hundred trillion boring ones. So in order to have a prayer of seeing anything interesting, you need to simply collide as many proton pairs as possible and hope for good luck. Further, if you want to make sure you understand the new phenomena you see, you need to wrap your detector like a sphere (or a cylinder in the case of ATLAS and CMS) around the entire collision point.

However, for different physics studies, you'd make different choices and nowhere at the LHC is this point made more apparent than in the LHCb experiment. Its main purpose is to study particles containing bottom quarks. These particles are called b-hadrons, where "hadron" means "particles containing quarks" and "b" reminds us that at least one of the quarks is a bottom quark. None of these b-hadrons are something you've likely to have heard of before, because they live only briefly—usually about a trillionth of a second.

And yet these short-lived particles can reveal interesting facts about the universe. Studying a class of b-hadrons containing a quark and antiquark, one of them of the bottom quark type, is beginning to shed light on the question of why the universe seems to be composed essentially entirely of matter. Further, it is thought that by precisely measuring the production and decay of b-hadrons that scientists might discover the Holy Grail of something new.

Because events in which b-hadrons are produced are relatively common (occurring in about 1% of all collisions), we don't have to collide nearly as many protons to study these kinds of collisions. Recall that during normal ATLAS and CMS running about 20 proton-proton collisions occur simultaneously. To accomplish this frantic rate, physicists must make very intense beams and put many protons in each.

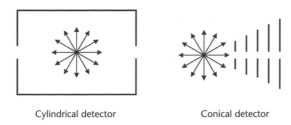

Cylindrical detector   Conical detector

Figure 4.17. A cylindrical detector envelopes the debris of the collision, while a conical detector only samples the debris. These techniques each have their merits, depending on the needs of the measurement. The breaks in the box (*left*) and the cone (*right*) show where material is intentionally removed to allow the beam to pass freely.

To study the production of b-hadrons, physicists still use proton beams, but these beams are much less intense than those needed by ATLAS or CMS. In these beams, usually only one pair of protons collides at a time. While the proton collisions still occur 40 million times a second, one collision at a time is much simpler than 20. In addition, the fundamental philosophy of the LHCb experiment is very different from that of ATLAS or CMS. ATLAS and CMS want to record and inspect the entire collision. If some new physical phenomenon is observed, the best way to understand it is to record all of the debris from the collision and inspect it all.

The LHCb experiment is mostly involved in studying b-hadrons. As long as the collision makes a b-hadron or two, that's a collision scientists might like to study. A corollary of this choice is that the LHCb experiment isn't concerned with recording all the debris from a collision. As long as the b-hadrons are recorded and measured accurately, that's good enough. Further, for technical reasons beyond the scope of this book, when b-hadrons are produced, they tend to be "near" the beam. "Near" means they tend to be produced in a narrow cone of about 40°, oriented on the beamline.

Consequently, the LHCb experiment has a much different geometry than ATLAS or CMS. Rather than a cylinder that envelopes the collision point, LHCb is cone-shaped and oriented to one side of the collision point; b-hadrons fly into the LHCb detector and are analyzed. Many particles created at the LHCb collision point entirely miss the detector. That's OK, as long as the b-hadrons are recorded. Figure 4.17 contrasts the ATLAS and CMS geometry to that of LHCb. With these introductory remarks in mind, we are now ready to look at LHCb in a little more detail. The LHCb detector is shown in Figure 4.18. You'll note a resemblance to Figures 4.14 and 4.16. LHCb looks like a slice of its larger counterparts.

Nearest the collision point is the Vertex Locator, or VELO, detector. The

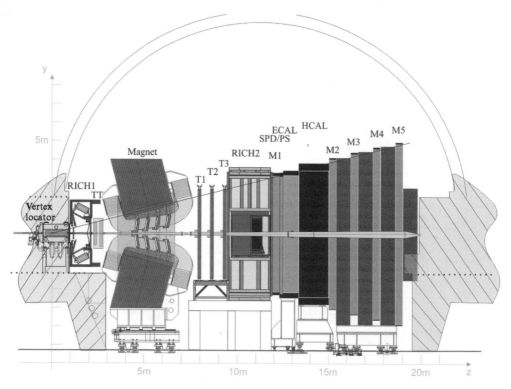

Figure 4.18. An overview of the LHCb detector, with major components identified. Note T1–T3 are the tracking chambers, M1–M5 are the muon detectors, and SPD/PS is a part of the energy-measuring calorimeter system. Figure courtesy CERN and the LHCb collaboration.

name stems from the detector's being designed to observe and measure the vertex caused by the b-hadron decay. The VELO detector is made of silicon, with strips about a 0.04–0.10 of a millimeter wide. Altogether, 172,000 strips of silicon make up this detector, compared with the 60 to 80 million detector elements in ATLAS and CMS. The VELO detector consists of 21 distinct layers of silicon, circular in shape and with a hole down the center. The detector has a passing resemblance to a series of CDs stacked and spread out over a meter or so. Figure 4.19 shows the basic mechanics of the VELO detector.

The designers of the VELO system made many clever engineering choices, but one is of interest to us here: Unless great care is taken, silicon is susceptible to damage by radiation. In normal operation, the beams would pass through the center of the hole in the center of the VELO disks. However, when the beam is being put into the LHC, it is larger and the danger of mis-steering it is greater. Thus the VELO is split into a left and right side, and the two sides can be re-

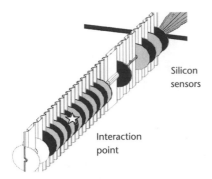

Figure 4.19. A close-up of the LHCb VELO system. Twenty-one disks of silicon surround the interaction point. Figure courtesy of CERN and the LHCb collaboration.

tracted during the moments when the beam is unstable and is likely to damage the silicon. In normal running, the VELO detector is positioned a scant 8 mm (0.3 inches) from the beam.

The next detector the debris encounters is one designed to help identify precisely which particles are passing through the LHCb detector. B-hadrons can decay into one particle or another, and precisely measuring how often the various possible decays occur is one of LHCb's goals. This detector is called a Ring Imaging CHerenkov detector (the designers used the more phonetic spelling of Cerenkov for their acronym), or RICH-1. The "1" is because there is a second RICH in LHCb (obviously RICH-2). The rest of the name comes from the fact that Cerenkov light comes out in the shape of a cone surrounding the particle's passage through the material. Depending on how fast the particle crosses the detector, the cone will be bigger or smaller. This cone of light hits detectors that convert the light to electricity and leaves a circular pattern. So, by measuring the energy of the particle and the size of the circle, one can frequently identify precisely what particle it is. It takes 200,000 photon detector elements in RICH-1 to properly reconstruct the circular patterns.

RICH-1 is followed by the trigger tracker (TT). This device is made of 180,000 strips of silicon arrayed in four layers. The layers are about 1.2 m (4 feet) high and about 1.5 m (5 feet) wide.

The TT is followed by a strong magnet, the job of which is to bend the path of charged particles traversing it. Once these paths are bent, the charged particles traverse the main tracking system. This system consists of 12 planes, grouped into three stations, each 4.6 m (15 feet) high and 5.8 m (19 feet) wide. This tracking system, depicted in Figure 4.20, consists of the inner and outer trackers. The inner tracker covers only 2% of the total area near the beam. This 2% of the area, while small, is where the particles are most concentrated and

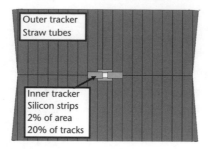

Figure 4.20. LHCb tracker, highlighting its two-component nature, with a small silicon detector at the heart, followed by a larger tracker consisting of strawlike plastic tubes. While the inner tracker covers only 2% of the detector area, it intercepts 20% of the tracks in a typical particle collision. Courtesy CERN and the LHCb collaboration.

captures a full 20% of the particles coming out of the collision. The inner tracker is made of 129,000 silicon strips, about 0.2 millimeters wide and 10 to 20 cm (4 to 8 inches) long. The outer tracker is made of long tubes, filled with a gas that ionizes when charged particles cross them. The outer tracker covers the bulk of the area (98%) with 54,000 tubes.

The tracking system is followed by RICH-2. Its purpose is similar to that of RICH-1: to help precisely determine the identity of the particles crossing it. RICH-2 contains 295,000 detector elements.

Just like the other big detectors, the LHCb tracking is followed by the calorimeters and muon-measuring systems. In most detectors, they occur in that order. But in LHCb, the calorimeters and muon system are intermixed, with the first layer of the muon system coming before the calorimeters.

The LHCb calorimeters are pretty traditional and are separated into an electromagnetic and hadronic part, both of the sampling variant. The electromagnetic calorimeter is made of lead to make the showers and plastic to measure the ionization. The hadronic calorimeter consists of layers of iron and plastic. Taken together, the calorimeters consist of a relatively modest 20,000 detector elements.

The final LHCb detector is the five-layered muon system, and it straddles the calorimeter, with one plane coming before the calorimeters and four after. Mostly, the muon system consists of large planes of wires that look like a very large harp. The wires of the harp are surrounded by a special gas that is ionized when crossed by charged particles. The wires carry the ionization energy out to waiting electronics. A small portion of the first layer of muon detectors consists of a technology that allows a more precise determination of the position of the muon's passage. The muon system's five planes each consist of about 25,000 different measurements to read out muon positions.

That the LHCb experiment contains something like 1% of the pieces of

ATLAS or CMS could be taken to mean that the LHCb experiment is inferior. Nothing can be further from the truth. For about 15% the cost of either of the other two larger detectors, LHCb will capture more than double the number of b-hadrons in any one collision than its larger brethren. This is because the LHCb detector was designed especially to study b-hadrons. When you are not designing for all measurements (and therefore have to be all things to all people), you can focus on your core game and perform better in your own little niche compared with your broader-scoped neighbors.

ALICE

The last of the big detectors at the LHC that we will discuss in detail is the ALICE (A Large Ion Collider Experiment) detector. Unlike the other three, which are optimized to study the debris of the collisions of proton pairs, ALICE is optimized to study collisions involving lead nuclei. Lead nuclei consist of 208 protons and neutrons, and thus these collisions are much more complex. The energy of the lead beams is immense. While for ordinary proton-proton collisions, the LHC can collide particles with an energy of 14 trillion electron volts of energy, in the case of lead collisions, the total energy is more like 1,150 trillion electron volts.

Given that the total energy involved in lead-lead collisions is 208 times larger than the proton-proton collisions, you might be wondering why people even bother with ATLAS and CMS. The reasons are straightforward. First, the energy in lead collisions is shared between many protons and neutrons. Thus the energy isn't as concentrated in lead collisions. It's like comparing the amount of heat in a room as compared to a flaring match. There's much more heat in the room, but still the match will burn you.

The second reason is that it is difficult to make intense beams of lead. Even working rather hard, the lead beams will only be about 10 million times less intense than the proton beams. Lead beams are so diffuse that even though they will cross as often as proton beams (40 million times a second), the lead beams will only collide about 8,000 times per second. When one insists that the lead nuclei hit head-on and not just a glancing blow, the collision rate will be a paltry 400 times per second, or about 100,000 times less often than the LHCb case and 2,000,000 times less often than in ATLAS and CMS.

With these relatively meager collision rate numbers, it might be a good idea to remind ourselves why we want to collide lead beams. It's because when a large number of particles are involved, one can observe different behavior. It's like at the zoo. As long as each species of animal is segregated in its own enclosure, the animals act in a particular way. But magically erase the cages and mix the animals together, then both types of animals, predators and prey alike, will exhibit behavior not seen before they were mixed.

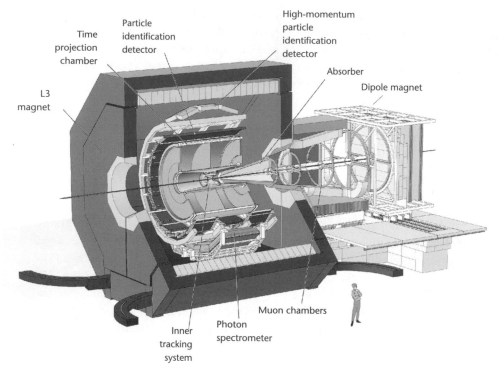

Figure 4.21. Cutaway view of ALICE, with major detector elements highlighted. Note the size of the person in the foreground. Courtesy CERN and the ALICE collaboration.

In a collision between two lead nuclei, 416 protons and neutrons get smashed into a small volume. With the intense heat of such a collision, it is perhaps unsurprising that many thousands of particles come out of the impact; in an extreme case, there can be as many as 20,000. Collisions between lead nuclei are a very messy business.

To study these highly complex interactions, the ALICE group designed and built 15 distinct detectors, using very different technologies. The overarching philosophy of the ALICE detector is not as simple as the cylindrical ATLAS and CMS detectors; nor is it like the conical LHCb detector. Indeed ALICE incorporates most of the tricks used in the other three detectors and a few other creative twists as well. Because the level of detail needed to thoroughly discuss the ALICE detector is so great, I've chosen to only sketch the high points of the design.

Figure 4.21 shows the ALICE detector. The detectors near the collision point are a series of cylinders, designed to track particles exiting the collision. Four distinct technologies make up these cylinders. They are a silicon tracker, a detector that utilizes ionization (the time projection chamber, or TPC), a detector that utilizes transition radiation, and a time of flight detector that measures how long it takes for a particle to travel from the collision point to the detector.

Following these detectors are two technologies that don't cover the entire cylinder. These two detectors are (a) Ring Imaging Cerenkov Counters and (b) lead tungstate crystals. Surrounding these detectors is a huge solenoid magnet. This detector has an abbreviated electromagnetic calorimeter (the lead tungstate crystals) and no hadronic calorimeter at all.

On one end of ALICE, one finds a conical style of detector, designed to measure muons. There are other detectors present in ALICE design, but we'll neglect to mention them in the interests of brevity. Following this short list of technologies, let's look briefly the various detectors, starting with the first layer, the inner tracking system or ITS, a silicon-based detector. The ITS sits in a cylinder about 1m (40 inches) long and about 0.9 m (34 inches) in diameter. The system consists of six layers of silicon, with each layer separated by 2.5 to 7.5 cm (1 to 3 inches). Not all layers are the full 1m (40 inches) long, with the layers closer to the beam being shorter. The inner two layers are silicon pixel detectors, while the outer ones are of the strip type. Altogether, the ALICE silicon detector is comprised of about 12.5 million detectors.

The next layer of ALICE uses ionization techniques. Charged particles cross the detector, which is mostly filled with gas. The charged particles ionize the gas by knocking electrons off the atoms. Strong electric fields guide the electrons to waiting electronics. By recording which electronics are hit and by measuring when these ionization electrons arrive, the path of the original charged particles can be determined.

This ionization chamber is the TPC mentioned earlier. It's a large hollow cylinder, with an inner radius of 0.75 m (2.5 feet) and an outer radius of 2.7 m (9 feet). The cylinder is 4.9 m (16 feet) long, and it contains 560,000 electronic channels.

The next detector utilizes transition radiation and is called the transition radiation detector, or TRD. It fits snugly around the TPC and itself is a hollow cylinder with an inner radius of just over 2.7 m (9 feet) and an outer radius of nearly 3.7 m (12 feet). The cylinder is about 6.7 m (22 feet) long, and the detector consists of about 1.2 million pieces. The purpose of the TRD is to distinguish between fast-moving electrons and the slower (and much, much more common) hadrons seen in collisions between lead nuclei.

The next detector is called a time of flight detector, or TOF. As its name suggests, it measures the amount of time it takes a particle to fly from the collision point to the detector. The basic idea is that one uses the particles' velocity to determine their identity. If two particles have the same energy, but one particle is much lighter than the other, the lighter particles will move faster and arrive more quickly. Now the differences in arrival times are pretty small. Taking two common hadrons, the difference in arrival time is about 3.5 billionths of a second.

The TOF detector is made mostly of gas and works like most ionization detectors. The difference is the electronics, which include the unusual feature of high-tech stopwatches that record the transit time. The TOF detector is very thin and is a shell about 0.3 m (a foot) thick and 8 m (24 feet) long, wrapped around the TRD. The TOF uses about 160,000 elements.

The next two detectors are odd. They don't cover the entire cylinder. Instead, they wrap around only a part of the cylinder and are rather short. In both cases, I have the mental image of a blanket on an elephant's back. The blanket doesn't fully cover the back, nor does it wrap around his stomach.

The reason these detectors don't cover the full cylinder are many. First and foremost is that by measuring what one sees in the small area covered by each of these detectors, one can project what's happening everywhere. Another consideration is cost. A final consideration is that the whole detector needs to fit inside the magnet (the next and final layer of ALICE). There wasn't enough room to put the two full cylinders in the remaining space.

The first of these two detectors is the high momentum particle identification detector or HMPID. The HMPID is made using Ring Imaging Cerenkov technology, basically similar to the RICH detectors in LHCb. As the name suggests, this detector is intended to determine the identity of highly energetic particles. This detector complements the TOF, which works best at low energy. The HMPID is composed of 161,000 detectors.

The second detector is called the photon spectrometer or PHOS. Like its name suggests, this detector is designed to measure photons. Consisting of 18,000 lead tungstate blocks, each about 2.5 cm (1 inch) square and 18 cm (7 inches) long, this detector has a passing resemblance to the CMS ECAL.

The ALICE detector doesn't contain a hadron calorimeter or a muon-detection system in its barrel. Instead, the final layer of ALICE is a large solenoid magnet. This magnet has seen service before. Recall that the tunnel in which the LHC accelerator is built used to house the LEP accelerator. The LEP accelerator hosted four experiments: Aleph, Delphi, L3, and Opal. The L3 magnet is now supplying ALICE's magnetic field.

The L3 magnet has a radius of 4.9 m (16 feet) and is 11.6 m (38 feet) long. The magnet supplies a relatively modest magnetic field, about 10,000 times that of the Earth and about one-eighth that of CMS. However, this lower strength is exactly what is needed for the kinds of particles ALICE will study.

The last major piece of ALICE is a muon detection system. This is not situated around the barrel, but off to one side, broadly similar to the geometry of

LHCb. Basically the muon system consists of five boxes, each box containing what look like two harps surrounded by gas. Muons cross the gas and knock electrons from the gas atoms, and the wires guide the electrons to waiting electronics. These five boxes of wires are combined with a magnet. The magnet bends the path of the charged particles and the planes of wires measure the amount of bend. From that, the muon's energy is measured. The magnet is about one-eighth as strong as CMS, and the muon detection system consists of 1.1 million detectors.

The ALICE detector consists of other subsystems as well, and these detectors play crucial roles in ALICE's discovery mission. However, these detectors have a more technical interest and a detailed description is omitted here. There are at least nine additional small detectors.

We have now learned something about the four major detectors at the LHC. These four detectors will commence collecting data in 2008 and will be at the forefront of physics research for at least the next two decades. It's impossible to predict which of the four detectors will make the crucial observation that reveals something entirely new.

Stay tuned!

Where We're Going

The Big Picture, the Universe, and the Future

Prediction is very hard, especially when it's about the future.

Yogi Berra

The journey into the atom and the realm of the small lead to outer space and takes us back in time to the birth of the universe. So far we've focused on the LHC and the kinds of scientific questions we hope that it will answer. However, there are exciting frontier questions for which the LHC will play a supporting role. To appreciate these important topics and the way in which the LHC is expected to contribute requires a different kind of story. This broader narrative illustrates the interconnectedness of the research frontier and underscores particle physics' deep connection to cosmology and the story of the birth and evolution of the universe.

Even though the LHC will have the highest energy of any accelerator in the world, it will not be the only one. It is expected that measurements made at other facilities will affect the search strategies at the LHC and vice versa. Further, even as the final construction phases of the LHC wind down, discussions are under way about the next accelerator. Given the amount of time it takes to propose, design, and build a particle physics facility, the planning for the future never stops. It is entirely likely that discoveries at the LHC will guide the design of the next facility.

In this chapter, we concentrate on these broader issues. We start with the recent epiphany that the universe is far more mysterious than we realized a mere decade ago. We know a lot about the quarks and leptons discussed in chapter 1 and can use that knowledge to explain everything in the universe that we can see. But "what we can see" is the operative phrase. It turns out that most of the universe is invisible. Indeed 95% of the matter and energy in the universe is something different from the ordinary matter with which we are familiar.

The Dark Side

Gravity is something about which we know a lot, at least in an astronomical context. The *New Horizons* space probe was launched on January 19, 2006, on a one-way trip to Pluto. Its goal is to pass within 16,000 km (10,000 miles) of the surface of Pluto and take the first detailed pictures of this ex-planet's surface. Including a flyby near Jupiter to assist in its speed, the *New Horizons* probe will travel about 5.3 billion kilometers (3.3 billion miles) and must hit a circle about 290 km (180 miles) in diameter in July 2015. Even though the probe carries a modest supply of fuel to correct any small deviations that may develop from the desired flight path, launching a spacecraft over such a large distance and hitting such a small target speaks volumes for the brilliance of Sir Isaac Newton and his law of universal gravity.

Newton was just 23 years old when he left London for the country to escape the Great Plague. The years of 1665 and 1666 might have been a very trying time for England, but it was a productive time for Newton. The invention of calculus and optics were both life defining achievements, but no matter how impressive those feats, it is his elucidation of gravity that begins our story.

Newton showed that the gravitational force between two heavenly bodies depended on four simple quantities. The first two are the masses of the two bodies, the third is the distance between the two, and the fourth is a constant that depends on the units being used and sets the scale for the strength of the force of gravity in the universe. This was a remarkable achievement. Using Newton's theory of gravity, combined with calculus, the motion of all planetary objects could be predicted. The magnitude of this achievement is demonstrated by the discovery of the planet Neptune and the (now) nonplanet Pluto because of subtle discrepancies between observations and predictions of Newton's gravitational theory.

Newton's law of gravity applies to more than astronomical bodies in our solar system. It also governs the rotation of galaxies and the motion of the galaxies themselves. The problem is that when Newton's ideas were applied to how galaxies rotated and the results of the calculations were compared with data that measured the actual rotation of galaxies, a glaring discrepancy was observed. So let's spend some time understanding both the calculation and the measurement.

The velocity of a star orbiting in a galaxy is governed by two factors: (a) how

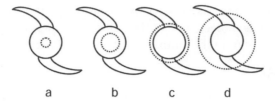

Figure 5.1. The dashed line illustrates various orbits of stars. As the radius increases from panel *a* to panel *d*, the orbit first quickly encompasses more mass, but as the radius of the star's orbit leaves the core of the galaxy and enters the galactic arms, the mass increase tapers off.

much mass exists within the star's orbit and (b) the distance between the star and the center of the galaxy. More mass and the star will orbit more quickly, larger distance and the star will orbit more slowly.

Figure 5.1 shows how the important parameters come into play. Most of the mass resides in the core of the galaxy, with the density of matter in the periphery being much lower. In Figure 5.1a, we see the orbit (denoted as a dashed line) is small and the fraction of the core's mass within the orbit is small. In Figure 5.1b, the radius has increased slightly but the amount of mass has increased greatly. In Figure 5.1c, the orbit is outside the core, so all the core's mass is within the orbit and you only get a relatively small amount of extra mass in the galaxy's arms. In Figure 5.1d, we continue to increase the radius but only get relatively modest gains in the mass inside the star's orbit.

When we take all factors into account, we find that the stars near the center of the galaxy should orbit slowly. As the distance from the center increases, the stars should orbit more quickly. This trend will continue until you get at such a large distance that you're leaving the bulk of the galaxy. Once you get to that distance, most of the mass of the galaxy is behind you. At these larger distances, the velocity of stars is predicted to drop off. This behavior is illustrated in Figure 5.2. For a detailed explanation of this behavior, refer to the suggested reading section at the end of the book.

With such a firm prediction in hand, astronomers could measure the rotation of galaxies and see how well the theoretical prediction agreed with the data. It was expected that, like other Newtonian successes, the data and prediction would agree beautifully. It was therefore a huge surprise when this expectation turned out to be so badly wrong.

The first study of this type was performed by Dutch astronomer Jan Oort, of Oort Cloud fame, in the late 1920s. He discovered stars in the Milky Way galaxy that were outside the main disk. These halo stars, as they were called, were shown to be orbiting the center of the galaxy. Oort found that these stars were moving more quickly than expected. One explanation was that the Milky Way

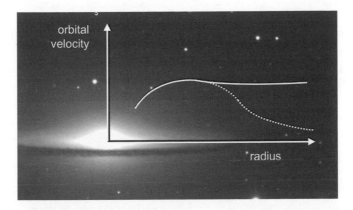

Figure 5.2. The orbital velocity of stars within a galaxy will increase quickly until the orbital radius is about the size of the galaxy. Beyond that, it is expected that the orbital velocity should decrease (dashed line). The measurement (solid line) doesn't agree with the prediction at large radii. Galaxy image courtesy Todd Boroson/NOAO/AURA/NSF.

had double the mass than was believed at the time. In astronomy, especially in that era, measuring something accurately to a factor of 10 was difficult, so if the mass of a galaxy had been determined incorrectly by a factor of two, it was not at all considered to be surprising and therefore people weren't terribly worried.

In the late 1950s, Louise Volders measured the motion of the stars in the galaxy M33 and came to a similar conclusion. And in 1970, Vera Rubin and W. Kent Ford Jr. published the first paper in what was (for Rubin) to be a career of measuring how galaxies rotated. In essentially all cases, the data and predictions agreed pretty well at distances near the center of the galaxies but disagreed more and more as the distance increased. Figure 5.2 shows a typical situation. Theory predicted that at very large distances the stellar orbital velocity should be less, but the measurement showed that above a certain distance, the orbital speed didn't change. This was very weird.

There are many possible explanations for this disagreement between data and calculation. The first is that the underlying premise was wrong and that Newton's laws were incorrect. It turns out that the accelerations experienced by stars orbiting galactic centers are very low. So possibly Newton's well-measured laws only work at "normal" accelerations and fail at lower accelerations. This idea was proposed by Mordehai Milgrom in 1983. While this explanation cannot yet be completely discounted, another possibility has become more popular. This other explanation was that there is more matter in galaxies than originally thought.

The way astronomers estimate a galaxy's mass is to figure out what kind of stars inhabit it. They then measure the brightness of the galaxy. By knowing the mass of stars of that particular brightness, you can figure out how many stars are necessary to explain the galaxy's brightness and thereby infer the galaxy's mass.

So critical to estimating a galaxy's mass is being able to see the matter within it. If some matter is invisible, it isn't taken into account. Because we use light (broadly defined to include other aspects of the electromagnetic spectrum, radio waves, for example) to see, then matter we don't see is called *dark matter.*

It should be emphasized that dark matter is merely a hypothesis and not the guaranteed truth often portrayed in popular writing. What is guaranteed is the observed matter and the fact that galaxies rotate more quickly than can be explained by Newton's laws. But the idea of dark matter is interesting, and it's fun to speculate what dark matter might be like and to see if there is any other evidence to strengthen the case to be made for it.

In 1933, Fritz Zwicky published his first paper in an obscure journal on the motion, not of stars in galaxies, but the motions of galaxies themselves. Galaxies don't wander the universe alone. Instead they clump together in large clusters of galaxies that might consist of hundreds or thousands of galaxies. A valuable mental image is a swarm of galactic bees buzzing around one another in a stately cosmic flight.

Zwicky's 1937 paper measured the velocity of the galaxies in the Coma cluster and estimated the mass of the galaxies in the usual way. The speed of galaxies and mass of the cluster are important variables and affect one another. Lots of mass and low velocities mean the cluster will eventually contract. Too much velocity and not enough mass and the galaxies fly away, as they are moving too quickly for gravity to pull them back. What Zwicky found was that the galaxies in the cluster were moving too quickly to be held there by the gravitational attraction of the visible mass. This additional evidence, coming as it did from a completely different direction, strengthened the case for the existence of dark matter.

So let's think for a bit just what dark matter might look like. Dark means "doesn't emit electromagnetic radiation." So dark matter could just be ordinary matter that is, well, dark. Heavy cool planets, black holes, burned-out stars, and large gas clouds are all examples of what is called *baryonic dark matter.* Baryons are a type of hadron; for our purposes they refer to protons and neutrons, the only stable baryons. So, essentially, "baryonic matter" is the ordinary matter with which you're familiar.

MACHO Bodies

Baryonic matter can exist as diffuse clouds of gas but those aren't terribly dark and can be observed using radio wave telescopes and the like. So that leaves

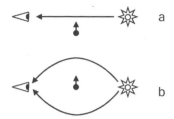

Figure 5.3. The light from the star going directly to the eye (*a*). However, when a mass comes between the star and the eye (*b*), light from more than one path is bent into the eye, making the star appear brighter. Note that the horizontal and vertical axes are not to the same scale.

us with "MAssive Compact Halo Objects," or MACHOs. As the name suggests, these are massive and small (compact), like black holes and burned-out stars. Further, they reside in the halo of the Milky Way galaxy, as the discrepancy between data and calculation observed by Rubin, Zwicky, and the others can be best explained by additional mass spread over the galactic halo.

Finding this dark matter is tricky. After all, by definition, you can't see it via any of the ordinary methods, and it only makes its presence felt via gravity. So astronomers had a clever idea. Since 1919, when Arthur Eddington successfully tested Einstein's theory of general relativity, physicists have known that gravity can bend light. This means that matter can act like a lens. As light passes by a mass, the light is deflected toward the mass. As shown in Figure 5.3, when a distant star sends light to your eye, the light that traveled directly from the star to your eye is what you see. However, if a mass passes between the star and your eye, you see a temporary apparent brightening as the mass gravitationally bends the light into your eye. So a good way to look for these compact objects is to simply watch distant stars and look to see if they ever brighten. If they do, then this can mean that a mass has passed between you and the star.

The brightness of stars can vary as a result of variation in how the stellar fire burns. However, that kind of variation affects each color differently. So to identify brightening from gravitational lensing, you have to look at the light from more than one color. If you see the same amount of brightening in all colors, you've observed a MACHO.

Because such gravitational lensing events are expected to be rare (these objects are compact after all), you need to look at lots and lots of stars and see if any of them brighten in the expected way. So scientists turned their telescopes on the nearby Magellanic Clouds as well as on our own Milky Way. Using modern telescopes, astronomers watched literally millions of distant stars, looking for the telltale brightening. And, in the spirit of scientific inquiry and competition, these observations were performed not by just one experimental group, but by

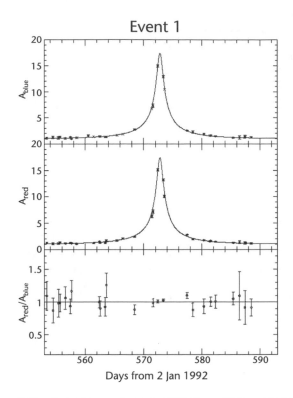

Figure 5.4. A gravitationally lensing event seen in 1994. Over a 10-day period, the light from this star brightened a lot. The top plot is for blue light, while the middle plot is for red. The bottom plot is the ratio of the two, showing that the two colors changed identically, the signature of a MACHO. Courtesy of the MACHO Project.

several. Each group competed with the others, cross-checking the other's results and making different observational choices, thereby extending the field's chances for success.

So what did they discover? Examples of these lensing events were observed. Such an example can be seen in Figure 5.4. Astronomers simply counted the number of these events and calculated the amount of mass these lensing events required. They could project their measurements to the entire galaxy and they came to one universal conclusion. There simply isn't enough mass of this form to explain the dark matter problem. So it was back to the drawing board.

It's Not Always Bad to Be a Wimp

If MACHOs could not explain the dark matter mystery, perhaps WIMPs could. WIMPs are "weakly interacting massive particles." Rather than dark matter being concentrated in the husks of burned-out stars, perhaps dark matter is instead dispersed across the cosmos, more like a gas. This candidate for dark matter would have to be a new form of matter, a particle thus far undiscovered.

Naturally, the words "undiscovered" and "particle" cause the ears of particle physicists to perk up. It's a conditioned reflex. The idea that a particle accelerator might crack the dark matter nut is a very attractive proposition to particle physicists. So if a new subatomic particle is the source of dark matter, what must its properties be?

To begin with, first and foremost it must be massive. After all, it is the unobserved mass in the universe that we're looking for. Further, since dark matter is invisible to our modern astronomical equipment, it must be electrically neutral. In addition, since it hasn't been observed, it can't experience the strong force. So we're looking for a stable heavy particle that experiences the gravitational force and possibly the weak force. Recall the supersymmetry discussion of chapter 2. The lightest supersymmetric particle (or LSP) just might fit the bill as a dark matter candidate.

A few questions remain. The first pair of questions is "Does dark matter of this form exist and how can we prove that it does?" The next question is "Even if dark matter does exist, how can we create it and manipulate it and thereby begin to understand its nature?" The answers to the first question will require observation "in the wild," as it were. Physicists will have to physically observe a new class of substance in the environment. Creating a new particle in the laboratory is insufficient. We need to see that it really is the answer to the dark matter question. However, while that is absolutely necessary, it is unlikely that studies "in the wild" will ever shed any light on the details of this hypothetical new matter. To study the details of a new particle, you need to be able to make it at will and study its properties. It is this aspect of the search for dark matter on which it is hoped the LHC will make a big statement.

Even if the LHC makes a new particle that would be an ideal candidate for dark matter, there is no guarantee that matter of that form exists in galaxies in large quantities. To establish that idea, physicists build small detectors consisting of a couple of kilograms of matter. The basic technique being used to search for dark matter is to just wait until a dark matter particle wanders through the equipment and bangs into an atom in the detector. Physicists painstakingly shield the equipment from outside influence, typically by locating the equipment deep underground. They cool the detectors to just barely above absolute zero, below $-273°C$ ($-459°F$). The reason the detectors are cooled is because if a detector isn't cool, the atoms in it are constantly vibrating. Further, the atoms are vibrating more than any expected signal caused by a dark matter particle's influence on the equipment. It is only through these extreme efforts to cool and shield the detectors that there is any chance of observing galactic dark matter.

However, while these observations-in-the-wild experiments are critical (and many are currently under way, with interesting performance-enhancing upgrades in the works), such studies will likely only reveal the mass of the dark matter particles and their frequency of occurrence. A successful determination of these properties would be an impressive scientific success, and there is a Nobel Prize in the future for the person or persons who manage to do it. However, experiments of this form will never be able to actually make this kind of matter. For that, you need an accelerator.

It's a race between the observational and accelerator studies to see who will find the first direct evidence for dark matter. The LHC is the only facility for at least the next two decades that will be able to create particles that *could* be dark matter. As with all research, there are no guarantees, but LHC researchers will be looking for particles that will fit the bill.

As a reminder, let's think about what sorts of particles the LHC might see that would look like dark matter. These particles would be massive, electrically neutral, stable, and not affected by the strong force. The particle would be invisible and, since it's stable (i.e., doesn't decay), we wouldn't see its daughter particles either. Thus the experimental signature would simply be energy disappearing in a particle collision. In short, we'll see any potential dark matter particle by not seeing it. The standard model particle, which is to say the particle we are sure we will see at the LHC, that behaves similarly is the neutrino. In short, if dark matter is observed at the LHC, it will look like a neutrino. And the way we know that it's a dark matter candidate and not just more neutrinos than expected is that we will see this excess only in collisions that are violent enough to have enough energy to make dark matter particles.

When we combine the results of all experiments looking for dark matter, we find that the amount of dark matter in the universe is much larger than ordinary matter. The amount of dark matter in the universe is about 50 times the glowing matter (stars, galaxies, etc.) we see in galaxies. When we include the dark clouds of intergalactic hydrogen, which account for the bulk of the ordinary baryonic matter and make up about 10 times the visible and glowing matter, dark matter still dominates. Dark matter seems to be about five times more prevalent than all ordinary matter. So the dark matter conundrum is a big deal. Not knowing the nature of 83% of the matter in the universe is embarrassing. However, there is a much bigger mystery in the universe. There appears to be an energy field in the universe that contains about double the energy and matter contained in dark matter and ordinary matter combined. This next mystery is called dark energy, and it's very mysterious indeed.

Matter Isn't the Only Thing That Can Be Dark

In 1998, two teams of astronomers and physicists discovered something obscure. The light from distant supernovae was too dim. It seems like such a little thing, and yet the consequences of that observation continue to ring through scientific circles and may well determine the fate of the universe. To understand the significance of this discovery we need to know how bright a supernova is and also how we know how far away it is. Those technical and somewhat arcane tasks have led us to a revolution in our scientific thinking.

One of the hardest things to do in astronomy is to know how far away an object is. A monkey gauges the distance to a branch to which it wants to jump by using its depth perception. Depth perception works because the monkey has two eyes, and the two eyes are separated by a short distance. However when large astronomical distances are involved, the binocular idea stops working. Even though we can treat the Earth at opposite sides of its orbit as two "eyes," this method only works for relatively close objects, say, a thousand light years. It's pretty useless for looking at distant galaxies.

Obviously we need a different method. To understand the crucial elements of these techniques, think about two people looking at the same candle with one person much closer to the candle than the other. The person closer to the candle will perceive it to be brighter.

The basic idea is simple and is illustrated in Figure 5.5. The light from a candle or any light source is uniformly emitted in all directions. As shown in the figure, at any particular distance, the light spreads out over the entire circle and the eye (or any detector) samples only a small fraction of the light. As the distance from the source to the eye increases, the circle gets bigger. Since the light is spread out over a bigger circle, the eye sees a smaller fraction of the candle's output.

This idea forms the basis of determining distance using apparent brightness. If you have a light of known brightness, you can predict the apparent brightness at any distance. Conversely, if you know how much light the candle is putting out and measure the brightness, you can determine precisely the distance between the eye and candle (or star and telescope in an astronomical context).

Measuring a star's apparent brightness is a piece of cake. The problem is to know how much light the object—in this case a star—is putting out. After all, different stars have different intrinsic brightnesses.

The study of the history of astronomy is littered with different astronomical bodies that have served this purpose. However, to see across the cosmos itself requires a really bright candle. Nothing less than the death of a star, called a supernova, will do.

Figure 5.5. A detector of fixed size (denoted by the thick line) covers a much smaller fraction of the total circle when placed far from the source of light. This makes the light appear dimmer, as you detect a smaller fraction of the light emitted.

A supernova occurs when a star dies in a spectacular thermonuclear explosion, during which a single star can briefly outshine its parent galaxy. There are different ways in which a star can be made to explode, but we're interested in a specific mechanism, or what is called a Type Ia supernova.

Type Ia supernovae are formed in binary star systems. One star is a white dwarf, and its gravitational field is so strong that it scoops matter from its celestial neighbor. As the mass flows down into the white dwarf, the star's mass grows and it gets hotter. Eventually, the mass and temperature get above a critical threshold, and the white dwarf blows up. Largely because the process is so similar in all cases, it is possible to know just how bright every Type Ia supernova is with an accuracy of about 10%, which in the astronomical world is "bang on."

Type Ia supernovae are rare and occur about once per century in each galaxy. But there are a lot of galaxies, and they all harbor Type Ia progenitors. Because we can determine precisely how bright they were to start with and we can measure very well the brightness we observe, we can calculate to impressive precision just how far away they are. This is an astounding achievement; to measure the distance to something half way across the visible universe and do it with an accuracy of about 10%.

Having one method to measure distances is incredibly valuable, but we are blessed. We have another method as well and can use both methods and compare them.

The second method for measuring distances uses the expansion of the universe itself. In 1929, Edwin Hubble found a relationship between the distance between our planet and a far away galaxy and how fast the galaxy was moving away from us. The more distant the galaxy, the faster it was receding. This was interpreted as evidence that the universe is expanding.

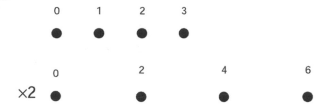

Figure 5.6. If you measure distances referenced from a specific point (point 0) and then everywhere double the distances, the farther away the point begins, then the farther (and therefore faster) it must move.

We can see how a simple expansion example illustrates this point. Look at Figure 5.6 and imagine yourself at point 0. There are points marked at 1, 2, and 3 units away. Now stretch everything to be twice as big in one second. This moves the 1, 2, and 3 points to positions 2, 4, and 6. Let's look how far each of these points moved. For 1 moving to 2, the movement was 1 unit. For 2 moving to 4, we have 2 units of motion and 3 moving to 6 yields 3 units of motion. Since the time it took for it to stretch was one second, the points at 1, 2, and 3 had to be moving with a speed of 1, 2, and 3 units per second. The more distant points move at a higher velocity compared with point 0.

The principle is the same in the universe. If you can determine the speed at which a galaxy is moving, you can determine how far away it is. This is precisely the experimental signature of the big bang and is a crucial reason why the big bang theory is nearly universally accepted.

Measuring the speed of a distant galaxy utilizes the same experience as watching a car race. One of the things you learn early on when watching a car race is that cars sound different whether they are approaching you or moving away. The sound of a race car passing you is "eeeeee-yooooo." It's a higher pitch (the "eeee" part) approaching you and a lower pitch (the "oooo" part) moving away from you. When the race car is the closest it gets to you, you hear the real car (the "y" part). High frequency means short wavelengths and low frequency means high wavelengths. Figure 5.7 shows this basic idea.

In astronomy, it's the same thing. Light consists of waves. In addition, stars consist dominantly of hydrogen, which emits characteristic hydrogen-only wavelengths of light. If a star is not moving with respect to your eye, you'll see the "right" color for hydrogen. However, if the star is moving toward you, the light wavelengths get scrunched up. Shorter wavelengths tend to be bluish, so we would call this light "blue shifted," as it would look bluer than it would if the star were stationary. Similarly, if a star is moving away from you, the light wavelengths will get stretched out. Longer wavelengths tend to be redder. Thus we

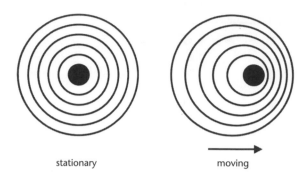

stationary   moving

Figure 5.7. An object (the dark circle) can emit waves. If it is stationary, the waves come out equally in all directions. If an object is moving to the right, then the waves will be shorter in the direction of motion and farther apart in the opposite direction.

say that the light from objects moving away from us is "red shifted." By measuring the amount the light is red shifted, we can determine the velocity. Finally, as we learned earlier, we can convert velocity into distance, as long as the expansion occurs at a known rate.

Each distant Type Ia supernova that was observed has also had its distance determined via red shift methods. Studies like these were independently published in 1998–1999 by two groups (the High-z Supernova Search and the Supernova Cosmology Project). When distances were compared using both methods, it was observed that supernovae appear to be farther way as determined by the brightness measurement when compared with the expansion measurement. This simple observation began a firestorm of cosmological rumination.

There have been many—an enormous number, in fact—explanations offered for these observations. While not all explanations have been received with equal enthusiasm, let me mention some of the more respectable alternative ideas.

- A long time ago, Type Ia supernovae were dimmer than they are now.
- Light is unstable and decays en route, or there is more matter or dust between here and the supernovae and the light is somehow being lost along the way.
- The universe is much larger than the small part we see. In fact, the entire universe is ringing like a bell from the aftershocks of the big bang. This ringing treats space like a Slinky, with some parts expanding and some contracting. We just happen to lie in an expanding spot.

Although these and other explanations remain viable with varying degrees of respectability, one explanation has begun to dominate cosmologists' thinking and that is the idea that the expansion of space is accelerating. The origin of this expansion is called "dark energy."

This is a thoroughly bizarre idea. Gravity, after all, is an attractive force. After the big bang, space was expanding wildly. This is the basis for Edwin Hubble's observation. But because gravity is attractive, we expect gravity to slow down the expansion of space. So far, there is nothing that suggests the expansion should increase.

In 1915, Albert Einstein formulated his theory of general relativity, which was experimentally verified in 1919 by Arthur Eddington. This theory describes gravity as the bending of space itself. The problem as Einstein saw it was that, given that gravity was an attractive force, his equations showed that the universe would eventually collapse. At the time, it was thought that the universe was eternal and unchanging, so Einstein added another factor to his equation. This new factor was effectively a repulsive gravitational force that would exactly balance out the traditional gravitational attraction, resulting in a static universe. Einstein was pretty pleased with himself.

With Edwin Hubble's discovery that the universe was expanding, the idea of the static universe died. Einstein removed his added repulsive factor, which was called the *cosmological constant,* from his equation and thereafter called it "his biggest blunder."

With the discovery in 1998 that the expansion of the universe is accelerating, the idea of the cosmological constant has undergone something of a renaissance. There are several ideas of how the cosmological constant might work, including one with the New Age sounding name of "quintessence" (although the idea is scientifically respectable). At any rate, the cosmological constant can be viewed as a constant-density energy field, filling the universe.

The concept of a constant-density energy field is counterintuitive, since as the universe expands, its volume also increases and therefore the energy from the cosmological constant increases too. But that's how it goes. So let's think a bit about the implications of this idea. Suppose there are two types of gravity. (Basically these two are one and the same and governed by Einstein's general relativity equations, but we'll treat them differently.) One type of gravity is our familiar variant, which is created by matter (both ordinary matter and the dark variant). This type of gravity is an attractive force and would therefore slow down the expansion of the universe. However, since this kind of gravity weakens as the distance between objects grows, the degree to which it slows the expansion of the universe is constantly decreasing as the universe expands.

The second kind of gravity is caused by dark energy, as embodied by the cosmological constant. The cosmological constant provides a constant-energy density and, more important, a constant repulsive force, which does not decrease as the universe expands. If dark energy provides a constant outward pressure and matter provides a steadily decreasing inward force as the universe expands, the outward pressure wins. Thus the expansion of the universe would accelerate. This is exactly what the comparison between the supernova and the red shift data indicates and is the primary reason why the dark energy hypothesis is so popular. Further, we now think that dark energy makes up 70% of the energy in the universe. Dark matter makes up about 25%, and ordinary matter makes up a mere 5%. Given that the dark component of the universe accounts for 95% of the universe's energy, it is natural that physicists are keen to winnow out its secrets.

So far in our discussion of dark energy, we've spoken entirely about cosmological and astronomical studies. However, this book is about the LHC. Where does Europe's new toy figure into the subject?

Dark energy is an energy field that permeates the universe. But nobody really has any clear idea as to its properties, except that you can calculate the amount of energy needed to agree with the astronomical observations. One of the LHC's most pressing goals is the observation and characterization of the Higgs boson. You may recall that the Higgs boson is the particle that makes up the Higgs energy field, which is the entity that gives all particles their mass.

I hope the phrase "Higgs energy field" caught your attention, because it did physicists who were thinking about dark energy. Could the Higgs field be the source of dark energy? Sure it could, except for one tiny thing. If you calculate the amount of energy the Higgs field must hold to provide particles their mass, it is much, much greater than the dark energy in the universe. Since we believe that the Higgs field or something like it must exist, there clearly is a mystery here. Perhaps when the LHC discovers (we hope!) the mechanism for electroweak symmetry breaking, the mystery might be solved. To add fuel to the fire, it is possible, using models incorporating supersymmetric principles, to exactly cancel out the effects of the Higgs field. This is an improvement, but still is a problem. Current theory finds the energy stored in the Higgs field to be much too high (by a factor of 1 followed by 50 or 100 zeros, depending on the details of the calculation). So if supersymmetry can exactly cancel out this huge energy, that's great. However, the problem remains that dark energy is not zero, just small compared with the Higgs field energy. So, although its successes are promising, even supersymmetry doesn't entirely solve the problem.

The bottom line is that studies of supersymmetry and the Higgs boson at the LHC will provide valuable guidance for cosmological questions. It has been long known that the particle realm and the story of the cosmos are intimately intertwined. The LHC will perhaps tell us the next chapter in that fascinating and never-ending story.

Other Accelerators

For the first part of chapter 5, we've concentrated on the amazing interconnections between the subatomic world and cosmology as a whole. However, the LHC is a particle accelerator, and it is with other accelerators that the LHC will mostly compete. To wrap up this chapter, let's spend some time talking about other accelerators, current and future, that will shape the particle physics world for the next decade or more.

As the LHC begins operations in 2008, there is only one operating accelerator with which it has to compete. This is the Fermilab Tevatron. The Tevatron is located at Fermi National Accelerator Laboratory (or Fermilab), located 64 km (40 miles) west of Chicago. Because the Tevatron commenced operations in 1983, it's tempting to think of it as a creaking old pile of equipment just barely holding together. However, nothing could be farther from the truth. For those sports fans out there, the Tevatron is a lot like Roger Clemens or Brett Favre, both athletes who continued to play long after their contemporaries retired and both who continued to teach younger competitors that outstanding talent and experience can beat just talent most of the time.

The Tevatron is a lower energy accelerator than the LHC. While the LHC will collide two beams of protons head on, with a total collision energy of 14 TeV (tera, or 10^{12}, electron volts), the Tevatron collides protons and antiprotons together with a collision energy of just 2 TeV. So the collision energy is much lower, only 14% that of the LHC. Further, the collision rate at the Tevatron is 50 to 100 times lower than what we expect to see at the LHC. While it's clear that these advantages make it inevitable that the LHC will eventually surpass the Tevatron, the simple fact is that when the LHC first turns on, it will not work up to design specifications. Whenever such a huge scientific apparatus is turned on for the first time, there will be inevitable teething pains. There may be difficulties bringing the accelerator to full energy and making the beams as bright as they eventually will be. In addition, the detectors that the LHC will host are brand new. Understanding the performance of anything that complex takes years. In contrast, the Tevatron hosts two large experiments, called CDF (for the Collider Detector at Fermilab) and DØ (which is the name of the detector's location at the Tevatron). CDF was originally commissioned in 1983, with DØ beginning operations in 1991. During the period from 1992 to 1996, both detectors recorded a great deal of data and simultaneously discovered the top quark. I was part of that excitement. During the period from 1996 to 2001, the Fermilab accelerator complex and both detectors underwent extensive upgrades. In 2001, we turned the equipment back on and have been in essentially constant opera-

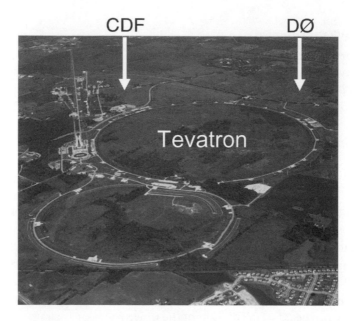

Figure 5.8. An aerial view of the Fermilab accelerator complex, 64.4 km (40 miles) west of Chicago, with detectors CDF and DØ indicated by arrows. Courtesy Fermilab.

tion since. Figures 5.8 and 5.9 show the Fermilab accelerator components and the two big detectors.

The upshot of all this is that both the accelerators and the detectors underwent massive refurbishment just seven years ago. Further, the physicists involved have had seven full years to understand any peculiarities of their equipment. It will be 2010 or even later before the LHC's physicists can realistically aspire to a similar understanding of their equipment.

All of this provides a window of opportunity for the Tevatron scientists. Even though the LHC was turned on for the first time in late summer of 2008, it will be 2010 or beyond before their understanding of their equipment will rival that of the Tevatron. During this period, the Tevatron still has a chance to scoop the LHC. And, as a Fermilab scientific staff member, I certainly hope that we do.

As of the fall of 2007, the Tevatron has authorization to continue to run through fiscal year 2009, which ends in October 2009. As I write, the lab management is attempting to make the case for an additional year of data taking, bringing us to the fall of 2010. Unless the LHC-based physicists get incredibly lucky, it will take them until then before their equipment is fully operational and understood at the level necessary to make competitive precision measurements.

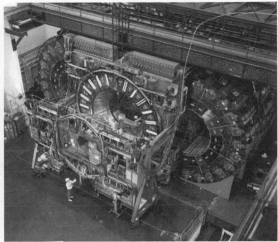

Figure 5.9. The detectors using the Fermilab Tevatron: DØ (*left*) and CDF (*right*). Courtesy Fermilab.

Further, the mass of the Higgs boson, if it exists, is unknown. However, the experiments using the LEP accelerator, once housed in the same tunnel now used for the LHC, were able to state that if the Higgs boson exists, its mass had to be higher than 114.4 GeV (giga, or a billion, electron volts), or about 120 times heavier than a proton. Different experiments and accelerators are able to explore different mass regions more or less precisely. As it happens, the Fermilab Tevatron is well suited to study the region just above the LEP experimental limit. While there is no doubt that the LHC can look there too, its strength is for potential Higgs boson masses slightly higher than that.

So the bottom line is that if the Higgs boson mass is only a little higher than the region excluded by the LEP experiments, then the Tevatron still has a chance to find it first. Given considerations of national pride and friendly interlaboratory competition, it is a certainty that Fermilab will concede nothing to the LHC and will continue to search until the Higgs boson is either found or it becomes apparent that the Tevatron is no longer competitive. No matter what, the Tevatron-based experiments will either win the race or go down swinging. There are worse ways for accelerators to close out their career.

What about That Thing in Texas?

During my frequent public lectures, I describe the successful scientific program of the Tevatron and the dreams of the LHC. Inevitably, someone asks about Fermilab's future. After all, when the LHC is up to speed, it will be hard for the Tevatron to compete. We'll get to Fermilab's future in a little while. But occasionally, someone will ask me, "Hey, what about that thing they were building in Texas?"

That "thing in Texas" as it's often called, was the ill-fated Superconducting Super Collider, or SSC. The SSC was intended to be America's version of the LHC. Proposed in 1983, the SSC was to be a huge accelerator 87 km (54 miles) in circumference, fully three times bigger than the LHC. The SSC was cancelled in the fall of 1993. While it's a little depressing to think about what might have been, the story of the SSC provides an important object lesson for future accelerator projects. So we'll briefly remember the story, without dwelling on it.

The SSC was to be America's answer to the LHC, although it was proposed earlier and had a projected turn-on date before Europe's LHC. Unlike the LHC, with its dual proton beams, colliding with a total energy of 14 TeV, the SSC would have collided a proton beam with an antiproton beam, just like the Fermilab Tevatron. However, unlike the Tevatron's (now relatively modest) collision energy of 2 TeV, the SSC was to have the astounding collision energy of 40 TeV, more than two and a half times higher than the LHC. Take that Europe! Everything is bigger in Texas!

Actually, the energy of the SSC had nothing to do with its being sited in Texas. (But try telling that to some of my colleagues from the Dallas area.) And, if truth be known, the SSC beams were designed to be only about a tenth as bright as the LHC. This was largely because of the need to manufacture antiprotons. It's just easier for the LHC to procure an unlimited supply of protons for its beams.

When the SSC was initially proposed, there was no official site in mind, except that it was to be built in the United States. Many states made bids to host the laboratory. At the time, I was affiliated with Rice University in Houston, Texas, and it was absolutely clear to me that the natural site for the SSC was at Fermilab, as it already had a tunnel dug for one of the smaller accelerators that would have fed the 87-km (54-mile) circumference final ring. Given that I firmly felt that Illinois was the SSC's best site, naturally a site just outside Dallas, Texas, was chosen. (I don't recommend that you let me pick the ponies for you at the track either.)

Waxahachie, Texas, was a green field site just as Weston, Illinois, was for Fermilab some decades earlier. Some of my ex-SSC colleagues tell me that a brown field site would be a more accurate description. Texas was chosen for a number of reasons; included among them was the state of Texas's promise to provide a very large sum of money for infrastructure and to build parts of the new accelerator complex. That it was a U.S. presidential election year with Texans George H.W. Bush and Lloyd Bentsen on opposing tickets may have played a role, too.

Groundbreaking for the SSC occurred in 1991. As the SSC was expected to be the future flagship of the U.S. particle physics program, it initially had appropriate funding. The original price estimate for the entire complex was in the neighborhood of 3 billion dollars. But that's when the trouble began. This price was always on the optimistic side. Delays and design changes rapidly escalated the cost estimate to more in the range of 8 to 10 billion dollars. With this increased cost, the SSC became a highly visible target for politicians wanting to show voters back home that they were being fiscally conservative. Plus, with the fall of communism in 1989, the baleful eye of the Soviet bear was gone, losing the SSC votes from politicians for whom the geopolitical game was a motivator. In 1993, Congress voted to terminate the SSC program.

So what lessons did the SSC debacle teach us? There were many. Probably one of the biggest was the realization that all such future projects would have to be internationally funded. No longer was it likely that the resources of a single nation state would foot the bill. The second lesson was that a good and honest engineering study must be done first, with a reasonably accurate price estimate. Large cost overruns are never good news. The final lesson to be learned is that scientists need to take the case for a new accelerator both to the public and to the politicians voting to fund it. While scientists did do this, it is clear that they could have done more. All of these lessons have guided the thinking of physicists designing an accelerator that is hoped to follow the LHC.

The LHC: The Future

Given that the LHC has not even been turned on yet, it sounds funny to be talking about an LHC upgrade. Yet it is the nature of modern particle physics accelerators that they have very long lead times. Thus thinking must start now.

The LHC environment will be hellish. Enormously intense beams of protons will circulate through the LHC ring. We expect that by 2014, some of the magnets making up the LHC will have absorbed so much radiation that they will no longer work very well. The detectors will also have absorbed a lot of radiation and will be in need of refurbishment. That's just six years from now.

By 2014, the LHC will have recorded about 100 times more data than the Fermilab Tevatron has since 1983 (with over 95% of that of Fermilab occurring since 2003 or so). With so much data recorded, another year of data taking at that rate and with that battered equipment would not be the desired path. So obviously on the time scale of six to eight years, the LHC will be ready for an upgrade and overhaul.

So what kinds of upgrades are being discussed? Before we describe the conversation, we need to always recall that if the LHC discovers some new physical phenomenon, then any talk of an upgrade will have to take that into account. But in the absence of such guiding knowledge, we can list some of the obvious ideas.

As with any accelerator, the LHC is defined by three things: beam type, energy, and brightness. Any upgrade of the LHC will likely keep the proton-proton collision scheme, but the other two parameters are open for discussion. For instance, the 14 TeV beam energy (two 7 TeV beams, hitting head-on) does not quite max out the LHC equipment. An engineering cushion has been built in to ensure stable operation. However, once the CERN accelerator scientists become comfortable operating the LHC, they might start to think about pushing the equipment a little harder. (Boys and their toys and all that, although women are well represented among CERN's accelerator scientists.) With current equipment, it is thought that the collision energy of the LHC can be raised to maybe 15 TeV, using the magnets now installed. A major energy increase, say a doubling to 28 TeV, will take new magnets and consequently a major R & D program.

The third attribute for a particle beam, the brightness of the beams, is more amenable to an improvement. The CERN accelerator scientists are already seriously discussing how to increase the LHC beam's brightness by a factor of 10 over the current design. This will be accomplished by putting more protons in the accelerator and by focusing the beams more.

The details of the LHC upgrade are impossible to state, as they are currently under discussion. However, an increase in brightness of a factor of 10 seems likely. If you recall from chapter 4, at current design beam brightness, each interesting (i.e., high energy) collision will be accompanied by 20 lower energy collisions. With the increase in beam brightness, the detectors will see 100 to 200 simultaneous collisions, depending on choices on how the increase in brightness will be achieved.

Because within the next few years the equipment in the major detectors (ALICE, ATLAS, CMS, and LHCb) will have been well worn, and, because the equipment was never intended to withstand the onslaught of the LHC's upgraded beams, detector upgrades will be in order. Without a clear idea of the plans for the LHC accelerator upgrade, it is impossible to accurately predict possible upgrades to the detectors. However, both accelerator and detector designers are currently discussing the options and deciding how to optimize the choices. The LHC is a premier scientific instrument and this upgrade will add a decade to its utility. Scientists at the LHC will be doing research well into the 2020s.

International Linear Collider

If the LHC is the near future, what of the more distant future? In the past, many frontier accelerators simultaneously operated, using different technologies.

Recently this variety was provided by Fermilab's proton-antiproton Tevatron, as well as the electron-positron machines at both CERN and SLAC, the Stanford Linear Accelerator Center. Currently, there is no near-term, frontier-energy electron-positron accelerator planned. The few electron-positron machines out there will be running at low energy and are specialty machines, studying bottom quarks to unprecedented accuracy. These accelerators could make new physics discoveries but only through increasingly subtle measurements.

And then, as an American, one might ask the natural question of the future of U.S.-based particle physics. With the shutdown of the Fermilab Tevatron in 2009 or 2010 or so, the United States, for the first time, would not have an energy-frontier accelerator. This doesn't mean that American scientists won't do frontier research. Indeed, Americans are well represented at the LHC. Of the 5,000 scientists and engineers working on the LHC experiments, more than a thousand of them are funded by U.S. institutions, making the United States one of the largest national groups working at the LHC. But an accelerator on U.S. soil would be better. On the other hand, the lessons of the SSC have not been lost on anyone. Any next accelerator will have to be international in character and not a U.S.-only endeavor.

Over the past decade or so, the international particle physics community has debated what accelerator is the natural one to build next. Many proposals have been floated, for instance the VLHC (Very Large Hadron Collider), an LHC-like device. The VLHC was to be the LHC on steroids, with a collision energy of 200 TeV, 15 times more energetic than the LHC and with a beam brightness similar to the LHC.

Another idea that was explored was a collider that would slam together two beams of muons. Muons are like heavy electrons, and this increased mass is attractive as muons would be immune to energy-loss problems inherent in accelerating electrons in a circle. (We'll describe those in a moment.) However, the big problem with muons is that they decay in a millionth of a second. So you'd have to make, accelerate, and collide these muons in an unprecedentedly short time. While some work continues on this idea, it has been mostly tabled for the moment. One attractive feature of a future muon collider is that it would make fantastically bright neutrino beams.

However, the idea that has gathered the most interest is a new electron-positron collider, and a most impressive idea it is. The last two high energy electron-positron accelerators were the LEP accelerator based at CERN and the SLC accelerator based at SLAC. Both accelerators were designed to copiously produce Z bosons. This meant that the collision energy was tuned to 0.091 TeV, or about 150 times smaller than that of the LHC.

The accelerator at SLAC was a linear accelerator, in which the electrons and positrons passed once through a long straight line of acceleration regions. (Hence the name *linear accelerator.*) The LEP accelerator, on the other hand, inhabited the tunnel that now houses the LHC. The 27-km (17-mile) circumference circular path meant that the path of the electrons and positrons was constantly being bent. And that's a problem.

When high energy electrons are accelerated in a direction perpendicular to the direction they are moving, they lose a little bit of energy through a mechanism called *synchrotron radiation.* The net effect is that the LEP accelerator had to constantly add energy to the beams that had been lost to the radiation. One can get an idea of how big a problem this was by comparing what fraction of the 27-km- (17-mile-) long accelerator was devoted to actual acceleration. For the LEP accelerators, this region was 600 m (2,000 feet) long. For the LHC, with its beam energies of 140 times greater than LEP, it is a mere 3 m (10 feet) long.

Because of synchrotron radiation, any larger electron-positron machines will have to be in a straight line. So this brings us to the basic idea of the next accelerator, the International Linear Collider, or ILC. The ILC is so-named because it embodies all of the most important principles we've learned over the years. It will necessarily be international in character, so no one country will bear the brunt of the full cost. It is linear for the reasons described above and because only by colliding beams head-on can you achieve the most energetic and desirable collisions.

As you read this, you must keep in mind that I can only paint the picture with very broad strokes. There is no detailed design yet and further the discussion is rapidly evolving. But there are ideas that are currently popular, and some choices have been made.

The ILC will fill a straight line tunnel about 27 km (17 miles) long. A beam of electrons and a beam of positrons will be aimed directly at one another and collide at the midpoint. Currently a collision energy of 0.5 TeV is the baseline design, with two beams of 0.25 TeV each. The accelerator designers have in the back of their minds the idea that an upgrade in collision energy to a full TeV would be natural.

You might be wondering about the collision energy. A seemingly modest 0.5 TeV is much smaller than the LHC's 14 TeV and is even lower than the soon-to-retire 2 TeV Tevatron. Doesn't this seem to be a step backward? The answer is yes and no, although mostly no. Recall that electrons are thought to be point particles, while protons are bags of quarks and gluons. Thus, while a proton may carry a lot of energy, each quark within it carries only a fraction of the energy. Since the collisions of interest to physicists occur between quarks, even at the LHC, the "interesting" collision energy is much lower than the total beam energy. Figure 5.10 illustrates these ideas.

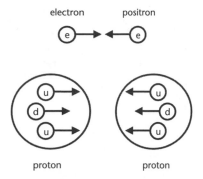

Figure 5.10. Electron-positron collisions are between pointlike particles and therefore all energy goes into the collision. In contrast, collisions between protons involve quarks (such as the up and down quarks indicated by *u* and *d*), none of which carry the proton's full energy.

Further, it is impossible to predict prior to the collision just what fraction of the proton energy a particular quark would carry. Therefore it is impossible to pick a particular desired collision energy. With an electron-positron machine such as the potential ILC, this is not an issue. Because electrons and positrons are pointlike, all the energy in the collision goes into the interesting part. Further, each and every collision will occur at exactly the same energy. Once you determine which energy gives the best results, you just tune your beams to that energy and study away.

So this brings up a point: Before any final choices are made in the design of the ILC, we have to get results from the LHC. Quite possibly (and hopefully!), the LHC will find something new. The ILC will then set up the beam parameters to look at just that thing. There is ample precedent for this in the past, for instance when the $Sp\bar{p}S$ (a proton-antiproton machine at CERN) found the Z boson and the LEP accelerator (an electron-positron machine, also at CERN) studied it with a precision that may never be surpassed.

Like most frontier accelerators, it is thought that the ILC will host two detectors. No decision has been made on ILC detectors, but there were four designs out there, which were just recently reduced to three. For a while, there were the United States, Asian, and European designs and one that came late to the game and is considerably more radical. Since the Asian and European designs were so similar, the groups decided in September 2007 to merge their efforts and produce one design that blended the best of both regional versions. Note that while I identified the detectors by geographical region, this isn't strictly correct, as all detectors are international in flavor. The regional designations just identify where the majority of the scientists who are working on that design call home.

One aspect of the ILC design that is a little controversial is how the detectors

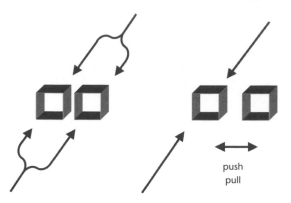

push
pull

Figure 5.11. Different ideas on how two detectors might share the ILC beam: A split beamline (*left*) or a single beamline with a push-pull mechanism (*right*).

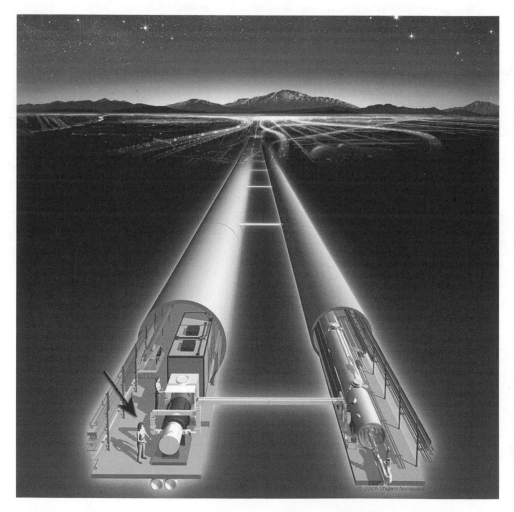

Figure 5.12. Artist's conception of what the ILC might look like. The accelerator is in the right-hand tunnel, with control electronics and power supplies in the left-most tunnel. Courtesy KEK.

Table 5.1 A comparison of several accelerators, current and future

Characteristic	Tevatron	LHC	SSC	LHC upgrade	ILC
Beams	proton antiproton	proton proton	proton antiproton	proton proton	electron positron
Energy (TeV)	2	14	40	14	0.5
Brightness relative to the Tevatron	1	100	10	1,000	200

Note: The SSC was never completed. TeV = one trillion electron volts.

will be placed in the beam. The most natural choice is to have two detectors side by side and to split the beams between the two as shown in Figure 5.11. However, this turns out to add a little under a billion dollars to the 10 billion dollar or so ILC price tag. So one money-conserving design change has been adopted (at least for now). The two detectors will be mounted on rollers and thereby be movable. Each detector will roll into the collision area and collect data for a while. Then the other detector will be swapped in for its turn. This idea is not without its detractors, and we'll see if the idea survives into the final design.

Figure 5.12 shows an artist's conception of what the ILC would look like. There would be two parallel tunnels, one containing the accelerator proper and the second housing power supplies, control electronics, and so on. The actual final design may vary from these early ideas, but it can be expected that this figure reflects something like the final design.

So where will the ILC be located? Given that the preliminary engineering work is not yet complete, that question is somewhat premature. After all, the decision to build it has not been made, nor will it be for several more years. Nonetheless, some regions have expressed an interest. Europe, of course, wouldn't mind hosting both of the world's great accelerators, but they will have their hands full with the LHC for the next decade or so. However, if the schedule for the ILC slips for several years, I expect that their interest to only grow.

Japan is a wealthy and technically adept country that has never hosted an energy-frontier accelerator. Further, they have been heavily involved in early ILC accelerator design. Some of my Japanese colleagues are very interested in working at an ILC based in the land of the rising sun.

Then there's the United States. With the turn-off of the Tevatron in 2010 or so, the United States will be without an energy-frontier accelerator for the first time. So naturally, American scientists would like to see the ILC be built here. (And that's not just because it's difficult in Europe and impossible in Japan to get a good hamburger in the cafeteria.) If it is built in the United States, scientists learned their lesson with the SSC debacle and are pushing for the ILC to be built at an operating laboratory. There are options, but Fermilab, just outside of

Chicago, is the favorite. Table 5.1 compares the various accelerators, present, future, and cancelled.

The ILC is currently just an idea, but it is consuming an increasing fraction of the particle physics community. No matter what comes after the LHC, the time to think about it is now.

Epilogue

The Large Hadron Collider began operations in 2008 and will dominate the energy frontier for at least a decade and probably longer. Nobody knows what mysteries of the universe it might reveal to us. We've discussed some of the bigger goals: the origins of mass, the lure of supersymmetry, and the chance to discover an entirely different level of matter, a new layer in the cosmic onion if you will. We've discussed how the LHC might contribute to the cosmic mysteries of dark matter and dark energy. The LHC might explain the mystery of the missing antimatter and it will certainly add to our knowledge of the first moments of the early universe.

For most people, solving these questions would be enough, but we particle physicists dream big. We are also looking to the LHC to help us explore the nature of space and time, particularly in understanding why we live in three spatial dimensions. It may even reveal additional unexpected dimensions. Experiments at the LHC could tell us something about quantum gravity, although that would be surprising. One thing that is for sure is that it will help us better understand those things that we already know something about.

The bottom line is that this is research and the LHC and all who work on it are explorers. Nobody can tell you what we'll find or indeed if we'll find anything at all. All the indications lead us to believe that there are thrilling new discoveries just around the corner. It is my colleagues' and my most fervent hope that we'll discover something new on this journey.

And, when we do, we'll send you a postcard from the quantum frontier.

Suggested Reading

No one book can explain all facets of any given subject in enough detail to satisfy every reader. Frequently a reader might want to know a lot more about a topic mentioned only briefly. Luckily, there are many good books written for a lay audience on a variety of physics topics and a reader can find that deeper and more thorough explanation that they are wanting. In this section, I try to recommend some of the better books available on the subject matter covered in this book.

It is difficult to organize such a list, as books often have multiple strengths. Thus I have chosen to list the books, with some commentary, and list for each which chapters you've read here those books overlap.

For general information about what we know about the universe and the particles it contains, I recommend *The Particle Garden* by Gordon Kane (Helix Books, 1996). It is a short book that describes very clearly what we currently know. It is written by a theoretical physicist, so it is light on experimental details. (Chapters 1 and 2)

For a more experimental treatment, my own *Understanding the Universe: From Quarks to the Cosmos* (World Scientific, 2004) is a better choice. This book is much longer and covers the history of particle physics, our current understanding of the standard model, accelerators and detectors, current mysteries, and particle physics links to cosmology. The treatment in my earlier book is aimed at a lay audience, but it is at a slightly more detailed level than this book. (Chapters 1, 2, 3, and 4)

For a light and breezy treatment of the history of particle physics, interspersed with a discussion of the universe as we currently understand it and culminating in a very short and nontechnical discussion of the Higgs boson, try Leon Lederman and Dick Teresi's *The God Particle: If the Universe Is the Answer, What Is the Question?* (Houghton Mifflin, 2006). Lederman's folksy style and Teresi's professional writing background are apparent throughout. (Chapters 1 and 2)

Gordon Kane's *Supersymmetry* (Perseus Publishing, 2001) is a book written ostensibly for a lay audience on the topic of supersymmetry and walks a very fine line between a lay audience and a nonmathematical treatment for a very junior scholar. For any serious first exposure to the topic, this book is a must. The reader should be aware that Kane is an ardent proponent of supersymmetry, so there is some merit to critics' comments that the book is not perfectly balanced and it leaves the reader with the impression that the existence of supersymmetry in the world is a more of a foregone conclusion than it actually is. (Chapter 2)

For a discussion of the important role that symmetry plays in modern particle theories, the book *Symmetry and the Beautiful Universe,* by Leon Lederman and Christopher Hill (Prometheus Books, 2004) is really quite nice. The idea of symmetry is sometimes daunting to the casual student of physics, and these authors do a good job of demystifying the topic. (Chapter 2)

For an accessible treatment about what we know that is somewhat more technical than what you've read here, try *Deep Down Things* by Bruce Schumm (Johns Hopkins, 2004). The reader should be aware that Schumm's book does break the taboo of popular literature, by occasionally including an equation. But these equations are used as spice rather than as an obstacle to understanding, and this choice will be welcome to all but the most math phobic. (Chapters 1 and 2)

The Charm of Strange Quarks: Mysteries and Revolutions of Particle Physics, by Michael Barnett, Henry Muhry, and Helen Quinn (Springer Verlag, 2002), is an unusual book. It covers the usual subjects, but the format is a mix of book, magazine, and textbook, with sidebars, column notes, and professionally drawn graphics. It has a vague similarity to the "X for Dummies" series (although it is entirely unrelated). It also is one of the few books that has any treatment of detectors. (Chapters 1, 2, 3, and 4)

Another unusual book is *The Particle Odyssey: A Journey to the Heart of Matter,* by Frank Close, Michael Marten, and Christine Sutton (Oxford University Press, 2002). This book can be described as a coffee table book, with extensive color photographs. It is a photo montage that includes history and future, even including some photographs of LHC prototypes. For those who need to see something to understand it, this is a very valuable book. (Chapter 1)

For those readers who like the stories of the history and the personalities as much as the physics, Martinus Veltman's *Facts and Mysteries in Elementary Particle Physics* (World Scientific, 2003) is a good choice. In addition to the usual descriptions of the physics we know, Veltman intersperses the text with one-page asides describing many of the colorful characters who have helped us understand our universe. As a Nobelist himself, Veltman is personally acquainted with many of these people and so many of his anecdotes have a firsthand flavor.

Veltman mentions accelerators and detectors, but the cursory treatment reflects his own high achievement as a first-rate theoretical mind. (Chapters 1 and 2)

While this book focused on the Large Hadron Collider, the last chapter flirted with the intriguing questions of dark matter and energy. With the recent discovery of what can be interpreted as dark energy, there has been an explosion of books on the subject. Dan Hooper's *Dark Cosmos* (HarperCollins, 2007) is an ideal introduction to the dark side of the universe for a reader who has no previous exposure. Hooper's book is light on details and paints with a hasty brush, but for a reader for whom the subject is entirely new, I highly recommend it. (Chapter 5)

A somewhat deeper treatment of the same material can be found in *Dark Side of the Universe: Dark Energy, Dark Matter, and the Fate of the Cosmos,* by Iain Nicolson (Johns Hopkins, 2007). Both dark matter and dark energy are described. (Chapter 5)

While the history of the discoveries of particle physics in the twentieth century is not a focus of this book, for a reader who is interested in the subject, I recommend the brilliantly written *The Second Creation,* by Robert Crease and Charles Mann (Rutgers University Press, 1996).

An interesting book that describes the discovery of the Z and W bosons and gives a real sense of the excitement and competition that goes along with a Nobel-bound discovery is *Nobel Dreams: Power, Deceit, and the Ultimate Experiment* by Gary Taubes (Tempus, 1986). This book is out of print, so you will need to get it from your library or an online out-of-print book source.

For a person who is interested in the history of CERN, it is hard to compete with *History of CERN,* volumes I, II, and III, by A. Hermann et al. (volumes I & II) and J. Krige (volume III), published in a three-volume paperback set by North Holland in 1996. These books are quite expensive and rare, so an interlibrary loan is your best bet to get access to them.

The astute reader will note that most of the suggested reading is related to the first two chapters, which is to say what we know and what our theories are looking for. Chapters 3 and 4, which describe the accelerator and detector principles, as well as details of the LHC complex, are uncommon. Partially this is because the LHC has not yet begun operations. I expect that this will change as time goes on. However, it also reflects an attitude among some that these are merely tools, and not as interesting as the discoveries they make possible. However, the history of science has always been an interplay between the discoveries and the equipment. It is impossible to fully appreciate why we believe the things we do if we don't understand the evidence. And one can never understand the evidence without an appreciation of the tools.

Finally, chapter 5 deals with the future and specifically one that has not been decided. Precisely what new facilities will be built will become apparent over the next few years. While what is written here is our thinking as of the summer of 2008, it is certain that the future as it unfolds will differ in some way. Your best chance for keeping current on these topics is to watch the popular science magazines.

For the more technically minded I recommend the journal article "General-Purpose Detectors for the Large Hadron Collider" by Daniel Froidevaux and Paris Sphicas in *Annual Reviews of Nuclear & Particle Science,* volume 56, pages 375–440, published in 2006. Note that this is a journal article, intended for other particle physicists, and is definitely *not* easy reading.

Web sites are always a dangerous thing to publish, because the World Wide Web is a fluid place and things change rapidly. However, there are some sites that are likely to exist for some time and would be helpful for the avid reader. They include the following:

The CERN home Web site: www.cern.ch/
The press office for CERN: http://press.web.cern.ch/press/
The ATLAS experiment: http://atlas.ch/
The CMS experiment: http://cms.cern.ch/
The LHCb experiment: http://lhcb.web.cern.ch/lhcb/
The ALICE experiment: http://aliceinfo.cern.ch/Public/
Particle physics news and images from across the world:
    www.interactions.org/cms/

Index

ECAL. *See* calorimeter(s): electromagnetic
Eddington, Arthur, 141
Eiffel Tower, 63
Einstein, Albert, 13, 30, 141, 149
electric current, 76
electric fields, 68–73; oscillating, 70–73
electric force. *See* force: electric
electromagnetic calorimeter (ECAL). *See* calorimeter: electromagnetic
electromagnetic force. *See* force: electromagnetic
electromagnets, 75–77
electron, 6–7, 14, 97
electron volt, 30, 70
electroweak symmetry breaking (EWSB), 25–35; alternatives, 35
electroweak unification, 25–35
elements, 5, 46
expansion of space, 147

Fermi, Enrico, 17
Fermilab (Fermi National Accelerator Laboratory), 62, 151
fermions, 37
fixed target mode, 84
floptwiddles, left-handed, 3
force, 18–21; electric, 11; electromagnetic, 11, 40; gravity, 11, 68; quantized, 19; relative strength, 19; strong, 11–12, 40; unification, 24, 39–41; weak, 11, 40
force-carrying particles, 20
force field, 19–20, 29
Ford, W. Kent, Jr., 139

galactic mass, 139–40
generations, fourth, 52. *See also* quark: generations
Gilligan's Island, 69
gluon, 20
gravitational lensing, 141
graviton, 20
gravity. *See* force: gravity

hadronic calorimeter (HCAL). *See* calorimeter: hadronic
hadrons, 97
haplons, 52
HCAL. *See* calorimeter: hadronic
heavy ions, 57–63; collisions of, 59
helons, 52
Higgs, Peter, 26, 28
Higgs boson, 26–35, 38, 153; decay, 30–33; evi-

dence for, 34; mass, 34; properties of, 29, 34; searches for, 34; supersymmetric, 42
Higgs field, 27, 150
High-Z Supernova Search, 148
hot air balloons, 65
Hubble, Edwin, 146, 149

ILC (International Linear Collider), 156–62; detectors, 159–61; future location, 161
International Linear Collider. *See* ILC
Intersecting Storage Rings. *See* ISR
ionization, 101–2, 107
ionization detectors, 109–10
ISR (Intersecting Storage Rings), 85

jet quenching, 59–60
jets, 31–32, 59–61
Joyce, James, 8
jug bands, metaphor for radio waves, 71
Jura Mountains, location of LHC in, 1, 88, 89–90

kindergarten teachers, 82
K mesons, 54

L3 magnet, 134
Large Electron Positron. *See* LEP
Large Hadron Collider. *See* LHC
Large Hadron Collider Beauty Experiment. *See* LHCb
Large Ion Collider Experiment, A. *See* ALICE
Large Hadron Collider Forward. *See* LHCf
lead glass, 111
lead tungstate, 116
LEP (Large Electron Positron), 34, 86–88, 134, 153, 158
leptons, 14–18; compositeness, 43–53; generation, 14; structure, 43–53
LHC (Large Hadron Collider), 150, 154; accelerator, 90, accelerator details, 88–95; beam dump, 93; beam energy, 93; beam structure, 80, 93; complex, 91; detector comparison, 122; lead beams, 94–95; magnets, 91–92; radiation damage, 155; RF cavities, 91; superconductivity, 92; types of magnets, 82; upgrade, 151–56; vacuum, 92
LHCb (Large Hadron Collider Beauty Experiment), 22, 62, 126–31; calorimeters, 130; muon system, 130; purpose, 25; RICH-1, 129; RICH-2, 130; silicon, 128; trigger tracker, 129–30; vertex locator (VELO), 127–29
LHCf (Large Hadron Collider Forward), 62, 125